AF691705

Fundamentals of Reliability in SiC Power Electronics

Ramin Katebi

Fundamentals of Reliability in SiC Power Electronics

by Ramin Katebi

© 2026 **DARYA SCIENTIFIC LLC**. All rights reserved.

No part of this publication may be reproduced, distributed, or transmitted in any form or by any means, including photocopying, recording, or other electronic or mechanical methods, without the prior written permission of the publisher, except in the case of brief quotations embodied in critical reviews and certain other noncommercial uses permitted by copyright law.

Publisher's Note:
The information in this book is for educational and research purposes. While the author and publisher have used their best efforts in preparing this book, they make no representations or warranties with respect to the accuracy or completeness of the contents.

ISBN: 979-8-9958805-0-9 (Hardcover)

Printed in the United States of America.
First Edition: April 2026

To Niloufar, Luna, and Prof. Nathan Weise

Preface

The transition of Silicon Carbide (SiC) from a laboratory curiosity to the backbone of modern power conversion has been one of the most significant shifts in semiconductor history. We have moved past the era where simply achieving high switching frequencies was enough to warrant a research paper. In today's landscape, the industry demands more than just performance; it demands **predictable longevity**.

As we witness the massive industrial shift from 150 mm to 200 mm wafers and the integration of SiC into everything from 800 V electric vehicle drivetrains to multi-megawatt grid stabilizers, the "reliability gap" has become the central challenge for the power electronics engineer. This book was born out of the necessity to close that gap.

Why This Book?

Most literature on power electronics treats the device as an ideal switch or a static data sheet entry. However, in high-power applications, the SiC MOSFET is a dynamic, living component influenced by electro-thermal interactions, packaging parasitics, and evolving failure modes like Bias Temperature Instability (BTI) and terrestrial cosmic ray susceptibility.

This text is structured to take the reader on a journey from the atomic to the systemic:

- **The Material Foundation:** Understanding the physics of $4H$-SiC and the complexities of the global supply chain that dictate device quality.
- **The Failure Physics:** A deep dive into the "why" behind the "magic smoke"—from short-circuit filaments to wire-bond fatigue.

- **The System Closure:** Moving beyond the die to explore how intelligent gate driving, advanced 3D packaging, and AI-driven health monitoring can turn reliability into a controllable software variable.

Who is it For?

Whether you are a graduate student trying to understand the transient behavior of a trench MOSFET or a veteran design engineer tasked with building a 3.3 kV stack that must survive twenty years in the field, this book provides the analytical tools and empirical data to guide your design choices.

The goal is not just to build converters that are smaller and faster, but to build systems that possess the "utility-grade" ruggedness required for a truly electrified future.

Ramin Katebi
San Jose, California
March 2026

Contents

Chapter 1

The State of SiC Power Electronics

1.1 Introduction

Silicon Carbide (SiC) has undergone an extraordinary transformation over the past three decades, evolving from an experimental substrate material into one of the most consequential enablers of modern power-electronic systems. Its wide bandgap of approximately 3.26 eV, high critical electric field approaching 3 MV/cm, and exceptional thermal conductivity deliver a combination of electrical and thermal robustness unmatched by conventional silicon technologies. These intrinsic advantages translate directly into device-level capabilities such as higher blocking voltage, lower specific on-resistance, faster switching, and higher maximum junction temperature. Analytical and compact models developed specifically for SiC MOSFETs further quantify how these material parameters map into conduction, switching, and short-circuit behavior over temperature.

System-level demonstrations consistently validate these benefits. Multiple traction-inverter studies report 3–10% drive-cycle efficiency improvements and significant reductions in cooling-system volume when SiC MOSFETs replace IGBTs in 800 V automotive platforms. Medium-voltage SiC technology is enabling new classes of high-power converters for rail, marine, and industrial drives. Renewable-energy converters—photovoltaic, wind, and energy-storage inverters—achieve notably reduced passive-component size and lower magnetic losses due to SiC's high-frequency operation. Aerospace and more-electric-aircraft (MEA) programs increasingly depend on SiC for its stability over temperature, radiation tolerance, and ability to withstand high dv/dt environments, with several recent prototypes flown in full-electric or hybrid-electric propulsion demonstrators.

Yet the advancements in SiC power electronics are not rooted solely in material superiority. Progress across epitaxy, substrate quality, MOS interface engineering, advanced device architectures, and high-reliability packaging technologies has been equally critical. Early SiC devices were limited by micropipe defects, high interface-trap densities, and packaging technologies adapted directly from silicon—none optimized for SiC's higher operating temperature or switching speed. Today, micropipes have been virtually eliminated, basal-plane dislocation densities reduced by orders of magnitude, and gate oxides optimized for long-term stability even under high electric fields. In parallel, packaging innovations such as advanced die-attach, low-inductance interconnects, double-sided cooling, and high-reliability ceramic substrates have enabled thermomechanical reliability far beyond traditional solder and Al_2O_3 solutions.

Equally important is the integrative nature of SiC technology: material physics, MOS interface reliability, device architecture, thermal management, packaging mechanics, gate-drive design, and system-level derating must all be co-engineered. This convergence is repeatedly emphasized across the SiC literature, including the references examined for this book, which spans device physics, packaging innovation, reliability models, automotive qualification, aerospace derating, and mission-profile stress analysis. Recent work on digital twins and physics-informed data-driven models further illustrates that SiC reliability can no longer be treated in isolation at the device level; it must be embedded into converter-level and system-level models.

This chapter establishes the foundation for the remainder of the book. It provides a comprehensive survey of SiC material physics, substrate evolution, epitaxial growth improvements, device architectures from diodes to trench MOSFETs, packaging requirements, and the major application domains that have driven SiC adoption. Later chapters build on this base to explore switching behavior, reliability modeling, module-level thermomechanics, qualification, and application-specific stress evolution.

1.2 Global SiC Supply Chain: From 150 mm to 200 mm Wafers

The SiC value chain is unusually "vertically sensitive": wafer defectivity, epitaxial uniformity, device yield, and module-level reliability are tightly coupled. In this ecosystem, upstream quality dictates downstream survival; a crystal-level defect in the initial boule inevitably propagates through epitaxy to become a terminal device failure. As a result, capacity expansion is constrained not only by wafer starts, but also by metrology throughput, epi reactors, high-temperature implantation/anneal capability, and qualification cycles that are longer than those typical in Si power electronics.

1.2.1 End-to-End Flow: Boule → Wafer → Epitaxy → Device → Module

A simplified SiC supply chain can be described by five serial stages: (i) physical vapor transport (PVT) boule growth and slicing, (ii) wafer finishing (CMP, surface preparation, inspection), (iii) epitaxial growth with tight doping and thickness control, (iv) device fabrication (implantation, activation anneal, oxidation/nitridation, metallization), and (v) packaging and qualification (die attach, substrate, interconnect, and screening). Because each stage has distinct yield loss mechanisms, the overall cost is dominated by compounded yield and by the slowest qualification bottleneck.

1.2.2 Why the Transition Matters: 150 mm (6") to 200 mm (8")

The industry shift from 150 mm to 200 mm wafers is driven by three economic levers:

1. **Die-per-wafer scaling:** For a fixed die size, 200 mm wafers enable a substantially larger number of die per wafer, improving fixed-cost amortization for batch-processing steps (such as furnace soaks or chemical baths) with weak diameter dependence.

2. **Cycle-time and tool utilization:** When toolsets (especially epi and high-temperature anneal) are designed for 200 mm, throughput per run improves; however, transitional periods often suffer

from mixed-diameter qualification and lower overall equipment effectiveness (OEE).

3. **Cost-down via learning curves:** 200 mm adoption forces process re-optimization (thermal budgets, uniformity, edge exclusion) and strengthens statistical process control (SPC), ultimately tightening parametric distributions and reducing outliers.

1.2.3 Yield and Defectivity: The Practical Constraint

For SiC, yield is not only a function of random point defects but also of extended defects—structural "scars" in the crystal lattice—that couple to electric field, leakage, and bipolar degradation. The 200 mm roadmap therefore requires lower defectivity, tighter epitaxial uniformity, and improved edge/termination process control. These drivers connect supply chain choices directly to device ruggedness; the detailed physics and defect taxonomy are consolidated in Chapter 2.

1.2.4 Cost Structure and Qualification Bottlenecks

The dominant cost contributors include boule growth energy/time, wafer finishing yield, epitaxial reactor utilization, and device-test/qualification fallout. Qualification for automotive and safety-critical deployments amplifies the economic penalty of statistically rare "tail failures" (latent defects that pass initial testing but fail prematurely in the field). These can force larger guard-bands, reduce usable operating area, or trigger aggressive additional screening.

1.2.5 Market Consequences: Capacity, Lead Times, and Design Choices

Supply-chain constraints propagate to module availability and can influence design choices such as discrete vs. module adoption, parallel die strategies, and conservative switching settings to broaden vendor options and ensure system-level reliability.

1.2.6 Outlook

The 200 mm transition is a statistical quality challenge as much as a diameter change. Progress is expected from improved in-line defect mapping, more mature epi control, and tighter integration between wafer suppliers, device fabs, and module manufacturers.

1.2.7 Geopolitics, Industrial Policy, and Supply-Chain Resilience

The migration to 200 mm manufacturing has turned SiC into a strategic supply-chain topic rather than a purely technical one. Because wafering, epitaxy, device fabrication, and module assembly are often distributed across different regions, geopolitical constraints can surface as yield and cost variability at the product level.

Regional manufacturing hubs: The USA, Europe, and China are each building vertically integrated SiC ecosystems, but with different constraints. US capacity expansion has emphasized domestic wafer and device fabrication, often supported by industrial-policy programs and defense/aerospace demand. Europe has focused on automotive-grade qualification, module manufacturing, and supply assurance for OEMs. China has accelerated capacity scale-up across substrates, epitaxy, and power modules to reduce import dependence and to support EV and renewable deployment at national scale.

Export controls and equipment constraints: Even when a region can grow boules and run epitaxy, access to leading-edge implant, metrology, and high-volume packaging equipment can limit ramp speed. As a result, the supply-chain bottleneck is frequently not the wafer diameter itself, but the *combination* of tool availability, process know-how, and statistically controlled defectivity.

Qualification and dual-sourcing as "hidden" cost drivers: Automotive and aerospace programs typically require dual sourcing, Production Part Approval Process (PPAP) documentation, and long-term change control. During the 150 mm→200 mm transition, qualification effort can dominate near-term cost even when die cost decreases, because new wafer lots and process flows must be re-characterized for breakdown statistics, oxide reliability, surge/avalanche robustness, and package life.

Implication for designers. Geopolitical risk manifests as second-source constraints, long lead times, and occasional specification shifts. A practical mitigation strategy is to design-in margin (thermal, voltage, and EMI), document derating assumptions, and maintain a validation plan robust to process-node and supplier transitions.

1.3 Material Foundations of SiC

1.3.1 Intrinsic Properties of 4H-SiC

Among all SiC polytypes, 4H-SiC has emerged as the dominant platform for power MOSFETs and diodes due to its favorable band structure, relatively high electron mobility, and excellent thermal properties. Table 1.1 compares several fundamental material parameters of silicon and 4H-SiC, illustrating the dramatic differences that enable SiC's superior device performance.

Table 1.1: Comparison of key material properties for Si and 4H–SiC.

Property	Si	4H–SiC	Impact on devices
Bandgap (eV)	1.12	3.26	Low leakage; high-temp operation
Critical field (MV/cm)	0.3	2.5–3.0	High-voltage; thinner drift layers
Thermal cond. (W/m·K)	150	370–490	Superior heat spreading
Electron mobility (cm^2/V·s)	1350	900	Lower mobility; mitigated by design
Saturation velocity (cm/s)	10^7	$2{\times}10^7$	Fast switching capability

1.3.2 Substrate and Ceramic Insulation

Ceramic substrates are essential for electrical isolation and thermal spreading. Traditional Al_2O_3-based direct bonded copper (DBC) substrates, while inexpensive, suffer from poor fracture toughness and limited cycling durability under SiC operating conditions. In contrast:

Substrate selection thus depends heavily on application domain (EV traction vs. industrial drives vs. aerospace) and required lifetime.

Panel-level fan-out packaging and embedded substrates have also been investigated to reduce module parasitics and improve mechanical integration, but they introduce challenges in warpage control and stress distribution.

1.3.3 Encapsulation and Insulation Reliability Under High dv/dt

SiC's fast switching produces dv/dt values exceeding 50 kV/μs. Standard epoxy encapsulants can experience partial discharge, erosion, and surface tracking under such electrical stress. These phenomena intensify with altitude (aerospace) due to reduced air pressure.

Modern insulation strategies include:

- Silicone gels with enhanced tracking resistance,
- Thin-film and conformal coatings with high breakdown strength,
- Field-controlled module layouts minimizing electric-field hotspots.

Thin-film and hermetic encapsulation approaches have demonstrated stable performance up to 250°C and under repeated thermal cycling, making them strong candidates for aerospace and high-density automotive modules. The interaction between encapsulation, partial discharge, and long-term insulation aging is an active area of research covered later in the packaging-focused chapters.

1.4 Broader Application Domains for SiC

The rapid commercialization of SiC extends beyond EV traction and renewable-energy inverters. This section provides a broader domain view.

1.4.1 Industrial Motor Drives

Industrial motor drives demand high reliability, long lifetime (>15 years), and resistance to elevated ambient temperatures. SiC enables:

- Higher switching frequencies,
- Improvements in motor acoustic noise and torque response,
- Smaller heat sinks and enclosure sizes.

These advantages are especially relevant for variable-speed compressors, pumps, and robotics, where efficiency at partial load and compact form factor translate directly into operational cost savings. SiC-based solid-state circuit breakers and motor control centers further extend the benefits by providing fast protection and flexible reconfiguration of industrial power systems.

1.4.2 Rail Traction and Transportation

Rail converters operate at megawatt-class power levels. SiC enables:

- Reduction in transformer size for medium-voltage converters,
- Lower switching losses in traction inverters,
- Improved efficiency at partial loads across varied duty cycles.

Hybrid Si/SiC and modular multilevel converter (MMC) based topologies are increasingly investigated for rail and large-vehicle traction, leveraging SiC for high-frequency stages and silicon or IGBTs for bulk conduction. Field trials in Europe and Asia have demonstrated SiC-based traction converters with significant weight, loss, and cooling-system reductions compared to Si-based solutions.

1.4.3 Data Center Power Delivery

Emerging applications in hyperscale data centers emphasize ultra-high efficiency at partial load. SiC MOSFETs in active front-end (AFE) converters and uninterruptible power supplies (UPS) have shown meaningful reductions in total cost of ownership through:

- improved efficiency at 10–40% load,
- reduced cooling infrastructure requirements,
- enhanced power density enabling rack-level consolidation.

Although many current deployments still use GaN at lower voltages, SiC is attractive for the medium-voltage stages that step down from utility distribution to data-hall distribution levels. Solid-state transformers and medium-voltage SiC rectifiers are likely to play a larger role as DC distribution gains traction in large-scale computing facilities.

1.4.4 High-Voltage DC (HVDC) and Grid Applications

Although HVDC historically relies on thyristor or IGBT technologies, SiC devices above 10 kV are emerging in research and early deployment. Their potential includes:

- lighter and smaller modular multilevel converters (MMCs),
- higher switching frequency enabling smaller filters,
- robust fault-handling capability under grid disturbances.

Prototypes of SiC-based MMC submodules and solid-state circuit breakers demonstrate reduced switching losses and improved thermal management over Si-based equivalents. These system-level experiments strongly motivate continued development of 10–15 kV SiC MOSFETs and hybrid SiC–IGBT switches.

1.5 Cross-Disciplinary Trends

SiC technology intersects materials science, semiconductor device physics, packaging mechanics, control theory, and system engineering. Several trends unify these disciplines:

- Increasing reliance on co-design—device structure, gate drive, thermal management, and packaging must be optimized together.
- Accelerated development toward 10–15 kV devices, driven by HVDC, rail, and grid modernization.
- Rapid increase in compact, high-frequency topologies (e.g., MHz LLC converters, SiC–GaN hybrid bridges).
- Strong emphasis on predictive reliability and digital twins for mission-profile forecasting.
- Growing use of converter-level health indicators such as gate-voltage dips, turn-off delay time, and on-state resistance shifts for online monitoring.

These trends reflect the maturity of SiC as a foundational technology rather than a specialty device. They also anticipate many of the topics explored in later chapters: active gate control, AI-enhanced monitoring, and multi-physics modeling.

1.6 Synthesis: The Evolution of SiC Power Electronics

The progression of SiC technology over the last two decades can be viewed as a tightly coupled evolution in four domains: materials, devices, packaging, and applications. While early adoption was driven mainly by Schottky diodes and first-generation planar MOSFETs, commercial competitiveness in automotive and aerospace markets demanded breakthroughs in epitaxy, defect control, trench architectures, oxide reliability, module integration, and system-level thermal design.

This interdependence is a defining feature of SiC technology:

- Improvements in epitaxy—most notably reductions in extended-defect densities—directly increased MOSFET yield and blocking capability.
- Device-architecture innovations (shielded trench gates, optimized cell geometries, multi-epitaxial drift layers) shifted the balance between conduction efficiency and oxide reliability.
- Packaging transitions from wirebond to copper clip and advanced die-attach enabled higher-frequency switching and survivability under thermal cycling and mission-profile stresses.
- Application-driven reliability requirements (AEC-Q101, DO-160, industrial derating curves) accelerated parallel research in MOS interface physics, time-dependent dielectric breakdown (TDDB) modeling, and predictive lifetime analysis.

Across the literature analyzed for this book, this co-evolution repeatedly emerges as the most important theme: SiC technology advances only when materials, devices, packaging, and application requirements are co-optimized.

1.7 Challenges and Opportunities Ahead

Despite remarkable progress, SiC technology continues to face unresolved challenges—many of which represent opportunities for the next decade of research and development.

1.7.1 Gate-Oxide Reliability Under Extreme Fields

Even with mature oxidation processes, the SiC/SiO2 interface remains more defect-prone than the Si equivalent, leading to lower channel mobility and susceptibility to bias temperature instability (BTI) and time-dependent dielectric breakdown (TDDB). Trench MOSFETs amplify these stresses but improve conduction dramatically. Further progress depends on:

- engineered oxide stacks (e.g., stacked oxides, nitrided oxides),
- optimized trench geometries (rounded corners, multi-step etch profiles),
- new materials (e.g., high-k dielectrics with interfacial layers).

Understanding how these design choices interact with real mission profiles, including repetitive short circuits, high dv/dt, and body-diode conduction, is central to future reliability improvements.

1.7.2 High-Voltage (>10 kV) MOSFET Development

Devices in the 10–15 kV class are already being demonstrated. These require:

- ultra-thick epitaxial drift regions (>80 μm),
- extremely low doping variation,
- suppression of high-field dislocation-induced leakage paths.

Such devices will enable SiC penetration into HVDC systems, medium-voltage drives, and grid-scale converters, potentially replacing multi-level stacks of lower-voltage devices. They also introduce new packaging, insulation, and test challenges, which are only beginning to be addressed in the literature.

1.7.3 Packaging for Harsh Environments

For aerospace, rail, and deep-industrial environments, thermal and mechanical stresses exceed the limits of today's packaging materials. Future packaging concepts may incorporate:

- ceramic-free power modules (direct-cooled),

- advanced dielectrics resistant to corona and partial discharge,
- metal-3D-printed heat-spreading architectures and embedded cooling.

Research on fan-out panel-level packaging, warpage optimization, and hermetic encapsulation hints at the direction of travel. The ability to integrate sensors and gate drivers inside the module further blurs the boundary between device, package, and converter.

1.7.4 AI-Driven Reliability and Digital Twins

Mission-profile–dependent degradation is increasingly modeled using digital twins—software models that replicate real module behavior under realistic thermal and electrical loading. This is particularly valuable for SiC devices, whose degradation pathways are sensitive to stress history and can be difficult to detect with simple offline tests. Machine learning is now being applied to:

- predict remaining useful life (RUL),
- infer degradation trends from in-situ measurements (e.g., gate voltage dip, turn-off delay),
- optimize derating strategies and active thermal control for long-term reliability.

These topics are explored in depth in later chapters of the book, especially in the context of automotive and aerospace qualification.

1.8 Conclusion

This chapter provided a comprehensive overview of the evolution of SiC power electronics, integrating insights from material science, epitaxial engineering, device architectures, packaging technology, and system-level applications. The interplay among these domains has shaped the SiC ecosystem: improvements in substrate quality enabled high-performance MOSFETs; advances in trench architectures required breakthroughs in oxide reliability; packaging innovations unlocked fast, high-frequency switching; and application-driven qualification frameworks shaped device and module design requirements.

The next chapter delves deeper into the physics of SiC MOSFETs, examining the drift-region design, MOS channel characteristics, switching behavior, and body-diode physics. Where Chapter 1 established the technological landscape, Chapter 2 provides the physical and mathematical foundation needed to understand the detailed reliability mechanisms examined later in the book.

Chapter 2

Fundamentals of SiC MOSFET Physics

2.1 Introduction

This chapter consolidates the *material-physics foundations* of 4H–SiC and the device-physics mechanisms that set the baseline for MOSFET behavior. Extended defect taxonomy (including BPD/TSD/TED families) is treated here in depth, linking growth quality to electric-field enhancement, leakage, and breakdown statistics. While the physics of defects and channel transport are covered here, their impact on long-term degradation and lifetime modeling is developed in Chapter 3.

2.2 SiC MOSFET Architecture and Schematic Layers

While the material properties discussed in Chapter 1 provide the foundation for Silicon Carbide (SiC) superiority, the realization of these benefits depends on the precise vertical and lateral architecture of the MOSFET. Unlike lateral silicon devices, power SiC MOSFETs typically utilize a vertical structure to maximize the breakdown voltage while minimizing the die area. This section details the physical layers and regions that constitute a modern trench-gate SiC MOSFET.

2.2.1 Vertical Structure: From Drain to Source

The SiC MOSFET is built upon a highly doped **N+ Substrate** (typically 4H-SiC), which serves as the mechanical base and the electrical drain contact. Moving upward, a thin **N+ Buffer layer** is often grown to mitigate defect propagation from the substrate into the active regions.

The most critical region for voltage blocking is the **N- Epitaxial (Drift) layer**. The thickness and doping concentration of this layer are engineered to support the target blocking voltage (V_{DS}). Due to the high critical electric field of SiC, this drift region can be significantly thinner than its silicon counterparts for the same voltage rating, which is the primary driver behind the low specific on-resistance ($R_{DS(on),sp}$) of SiC devices.

2.2.2 Device Geometry and Active Regions

The top portion of the device contains the active regions responsible for switching control:

- **P-Well (Body Region):** Implanted into the drift layer, this region forms the PN junction that blocks voltage in the OFF state. It also houses the channel.

- **N+ Source Regions:** Highly doped regions where electrons are injected into the device when the channel is formed.

- **P+ Shielding:** In trench architectures, specialized P+ regions are often placed beneath the trench to protect the gate oxide from high electric field peaks during the OFF state, a critical reliability feature unique to SiC.

2.2.3 Layered Composition and Schematic Representation

The physical relationship between these layers is illustrated in Figure 2.1. This schematic represents a *trench-gate* architecture, where the gate electrode is etched into the SiC surface to increase cell density and eliminate the JFET resistance found in planar designs.

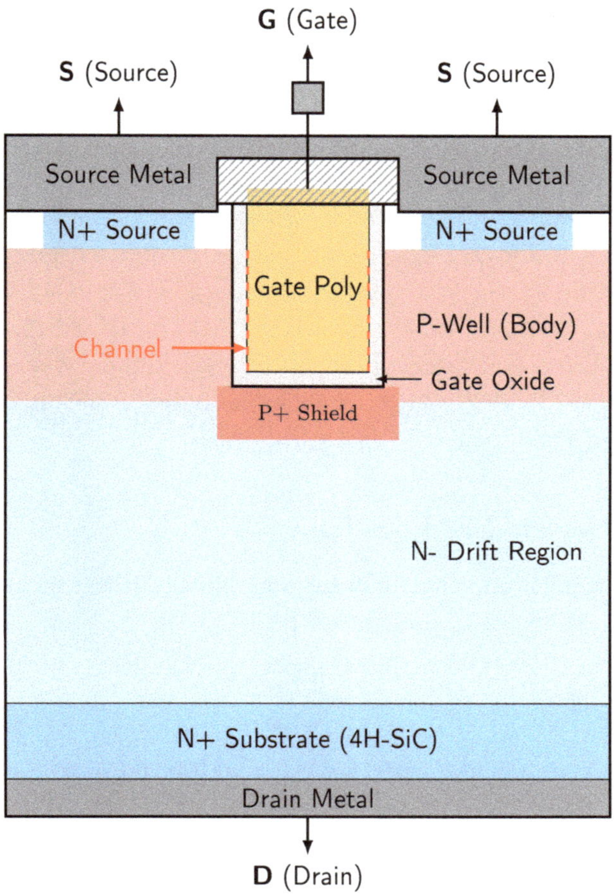

Figure 2.1: Cross-sectional schematic of a Trench SiC MOSFET showing the vertical layer stack and active regions.

2.2.4 Basic Principle of Operation

The SiC MOSFET operates as a voltage-controlled switch. In the **OFF state** ($V_{GS} = 0V$), the P-well/N-drift junction is reverse-biased, and the drift region supports the full drain-to-source voltage.

In the **ON state**, a positive voltage applied to the gate ($V_{GS} > V_{th}$) creates an electric field that attracts electrons to the SiC/SiO_2 interface along the sidewalls of the trench. This forms an **inversion layer** (channel), providing a conductive path for electrons to flow from the N+ source, through the channel, into the N- drift region, and finally to the drain contact. The physics of this electron transport, including the impact of interface traps and field scaling, are discussed in the following sections.

2.3 Drift-Region Physics and Breakdown Behavior

The drift region is the dominant contributor to specific on-resistance in high-voltage SiC MOSFETs. As the voltage rating increases, the drift layer must be thicker and more lightly doped to withstand avalanche breakdown. Because the critical electric field in 4H–SiC can exceed $3\,\mathrm{MV\,cm^{-1}}$, the drift layer can be dramatically thinner than in silicon for the same blocking voltage, enabling an order-of-magnitude reduction in drift resistance.

2.3.1 Electric-Field Scaling in 4H–SiC

For a one-sided abrupt junction, the approximate relationship between breakdown voltage V_{BR} and critical field E_{crit} is:

$$V_{\mathrm{BR}} \approx \frac{E_{\mathrm{crit}}^2 \varepsilon_{\mathrm{SiC}}}{2qN_D}, \tag{2.1}$$

where N_D is the drift-layer doping, q is the elementary charge, and $\varepsilon_{\mathrm{SiC}}$ is the permittivity of SiC. Because SiC possesses a critical electric field strength nearly ten times greater than that of silicon, the drift-layer doping can be increased by nearly an order of magnitude for the same voltage rating without compromising the device's breakdown strength. This higher permissible doping directly reduces the drift-layer resistivity compared to Si devices.

In real devices, field plates, junction-termination extensions (JTE), and beveled or trenched terminations are used to spread the electric field and approach the ideal 1-D scaling. Many experimental studies in the literature show that advanced termination schemes are particularly important for 3.3 kV and 10–15 kV classes, where edge effects dominate breakdown behavior.

Table 2.1 summarizes representative drift-region parameters for commercial voltage classes.

These drift geometries determine conduction losses, switching speed, and avalanche robustness. Precise control of doping uniformity—typically within ±5% across a 150–200 mm wafer—is necessary to achieve narrow breakdown distributions and predictable ruggedness in power modules.

Table 2.1: Representative drift-region thickness and doping for SiC MOSFETs across voltage classes (approximate values).

Voltage	Thick. (μm)	Doping (cm^{-3})	Notes
650 V	6–10	5×10^{15}	Low $R_{\mathrm{on,sp}}$; channel dominates
1200 V	10–15	$(2\text{–}4)\times10^{15}$	Standard automotive class
1700 V	18–25	$(1\text{–}2)\times10^{15}$	MV drives, fast chargers
3300 V	40–60	5×10^{14}	Traction and MV converters
10–15 kV	80–120	$(1\text{–}5)\times10^{14}$	HVDC and grid applications

2.3.2 Impact of Crystal Defects on Breakdown

SiC drift regions contain extended defects such as threading screw dislocations (TSDs), threading edge dislocations (TEDs), and basal-plane dislocations (BPDs). Their impact includes:

- **TSDs** produce localized field enhancement, lowering V_{BR} and increasing leakage.
- **TEDs** contribute to soft breakdown behavior and elevated off-state leakage at high temperature due to trap-assisted tunneling.
- **BPDs** have limited impact on purely unipolar drift conduction but play a dominant role in bipolar degradation of the body diode (Section 3.5).

Statistical breakdown measurements in several studies show a strong correlation between TSD density and the Weibull distribution of breakdown voltages, especially for 3.3 kV and 10 kV devices. Devices fabricated on low-TSD wafers exhibit steeper Weibull slopes and higher median breakdown voltages, simplifying derating decisions for traction and grid applications.

In addition to line defects, point defects, stacking faults, and epitaxial thickness variations also impact breakdown and leakage. Work on avalanche ruggedness shows that devices with optimized epitaxial design and low dislocation density can sustain substantial unclamped inductive switching (UIS) energy without catastrophic failure, but the margin shrinks at higher temperature and for trench-based architectures.

2.3.3 Polytypes, Crystal Structure, and Defect Evolution

SiC exhibits over 200 polytypes, of which 4H, 6H, and 3C are the most widely studied. The 4H polytype provides the optimal balance of mobility and bandgap for unipolar power devices. Historically, the limiting factor in SiC technology has not been intrinsic physics but crystal defects, especially micropipes and basal-plane dislocations (BPDs).

Earlier SiC wafers suffered from micropipe densities exceeding hundreds per cm^2, making high-voltage devices nearly impossible to fabricate with acceptable yield. Modern SiC substrates have eliminated micropipes entirely, and threading screw dislocations (TSDs) have been reduced to 10^2–10^3 cm^{-2}. Epitaxial growth processes have improved uniformity, reducing doping variation, mitigating carrot defects, and limiting BPD propagation during high-current body-diode conduction. These improvements have enabled reliable devices at 1.2, 1.7, and 3.3 kV ratings, and form the foundation for emerging 6.5–10 kV devices.

Remaining defects still influence reliability in nuanced ways. BPDs can transform into stacking faults under bipolar conduction, degrading body-diode performance. Threading screw dislocations (TSDs) influence breakdown voltage statistics and can act as leakage paths in high-field regions. Threading edge dislocations (TEDs) can modulate leakage currents under high electric fields and impact device-to-device variability. These mechanisms receive deeper analysis in Chapters 2 and 3, where their interaction with mission profiles, avalanche events, and short-circuit stress is quantified.

2.3.4 Substrate Defect Evolution

Three primary dislocation types affect SiC substrate performance:

- **Threading screw dislocations (TSDs):** strongly correlated with breakdown behavior and local field crowding.
- **Threading edge dislocations (TEDs):** affect leakage and local electric-field distribution.

- **Basal-plane dislocations (BPDs):** trigger stacking-fault propagation under bipolar conduction.

Extensive mapping of defect density versus die performance shows that high-voltage devices (> 3.3 kV) are particularly sensitive to TSD concentration. Meanwhile, TEDs are implicated in reverse leakage, especially at elevated temperatures and under radiation stress. The steady decline in defect densities is one of the most significant contributors to SiC's commercialization: substrates with micropipe densities of hundreds per cm^2 in the 1990s are now routinely manufactured with zero micropipes, and TSD densities below $10^2\ cm^{-2}$ are common.

The defect landscape also interacts with radiation and single-event phenomena. Studies in nuclear and space environments indicate that local field enhancements around dislocations can act as starting points for heavy-ion-induced leakage and single-event burnout. These results motivate conservative derating and careful screening for space and high-altitude applications, discussed later in the reliability chapters.

2.3.5 Drift-Region Resistance & Baliga's Figure of Merit

The specific on-resistance of the drift layer in a unipolar device can be approximated as:

$$R_{\text{on,sp,drift}} = \frac{4V_{\text{BR}}^2}{\mu_n \varepsilon_{\text{SiC}} E_{\text{crit}}^3}, \tag{2.2}$$

where μ_n is the electron mobility in the drift region, ε_{SiC} is the permittivity of SiC and V_{BR} is the break down voltage. This expression captures Baliga's unipolar figure of merit and highlights SiC's advantage:

- The critical field E_{crit} appears with the *third power* in the denominator.
- For the same V_{BR}, the drift resistance in SiC can be reduced by $50 - 100$ times compared to Si.

Figure 2.2 illustrates the theoretical versus practical drift-layer resistance across voltage classes, comparing ideal scaling with representative published device data.

The gap between theory and practice is primarily filled by channel resistance, JFET constriction, contact resistance, and layout con-

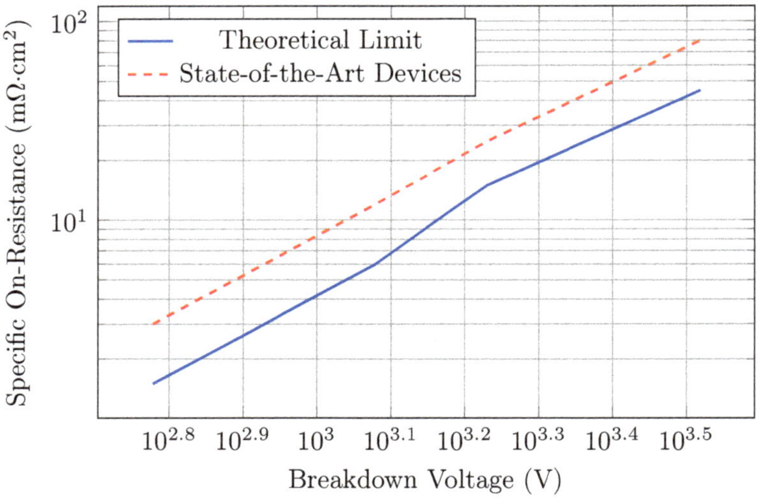

Figure 2.2: Representative theoretical and practical drift-layer resistance scaling for SiC MOSFETs. Practical devices remain above the theoretical limit due to channel mobility limits, JFET resistance, and contact resistances.

straints. Understanding and minimizing these contributions is the subject of the next section.

2.4 MOS Channel Physics: Interface Traps and Mobility

Although SiC's drift region offers enormous advantages, the MOS channel has historically limited MOSFET performance, particularly at low and medium voltages (650–900 V classes). The SiC/SiO_2 interface contains a high density of interface traps (D_{it}) that capture carriers and reduce *inversion-layer mobility*—the ease with which electrons can move through the thin conductive path formed under the gate oxide. As explored in Section 2.4.2, these traps act as electrical obstacles, turning a theoretically smooth channel into a highly resistive one.

2.4.1 Interface Trap Characteristics

Interface traps in 4H–SiC MOSFETs typically fall in midgap and near-conduction-band energy ranges, with densities:

$$D_{it} \approx 10^{12} - 10^{13}\ \mathrm{cm}^{-2}\mathrm{eV}^{-1},$$

far higher than those found in silicon MOSFETs.

To resolve these distributions and their associated time constants, several experimental techniques are utilized;

- High-low C–V: This method compares capacitance-voltage curves measured at a high frequency (e.g., 1 MHz), where traps are too slow to respond, against low-frequency or quasistatic curves where traps can follow the signal. The "gap" between these curves represents the interface state capacitance, allowing for D_{it} extraction. In SiC, this is often performed at elevated temperatures to "activate" deep-level traps that are otherwise frozen at room temperature.
- Conductance Method: Considered more sensitive than C–V techniques, this method measures the energy loss (conductance, G_p/ω) as traps exchange carriers with the conduction or valence bands. By plotting conductance against frequency, the peak of the curve directly identifies the trap time constants (τ_{it}), providing a map of how "fast" or "slow" the traps are at specific energy levels.
- Stress-Induced Leakage Current (SILC): This technique monitors the integrity of the gate oxide by measuring minute increases in gate leakage current following electrical stress. SILC is a sensitive indicator of the generation of new traps and defects within the oxide bulk, serving as an early-warning signal for potential gate-oxide breakdown.

The consequences include:

- Reduced inversion-layer mobility (often only 20–40% of the bulk mobility).
- Threshold-voltage hysteresis due to slow capture and emission dynamics.
- Bias temperature instability (BTI), with positive BTI (PBTI) under positive gate bias and negative BTI (NBTI) during body-diode conduction or negative gate bias.

Several studies in the literature show that traps with time constants in the millisecond to second range strongly influence dynamic on-resistance and transient behavior under pulsed-gate conditions.

2.4.2 Mobility Models and Impact on Performance

The inversion-layer mobility μ_{eff} in 4H-SiC MOSFETs is significantly lower than the bulk mobility (typically 10–50 $cm^2/V{\cdot}s$ versus $\approx$ 900 $cm^2/V{\cdot}s$). This effective mobility is modeled using Matthiessen's rule to account for the various scattering mechanisms present at the SiC/SiO_2 interface:

$$\frac{1}{\mu_{eff}} = \frac{1}{\mu_{Coul}} + \frac{1}{\mu_{phonon}} + \frac{1}{\mu_{surf}} \tag{2.3}$$

where:

- μ_{Coul} captures Coulomb scattering from charged interface traps (D_{it}) and fixed oxide charges. This term dominates at low-to-moderate electric fields and is the primary cause of the low mobility observed in 4H-SiC devices.
- μ_{phonon} accounts for lattice (phonon) scattering. This mechanism becomes more prominent as temperature increases ($\mu_{ph} \propto T^{-2.5}$), leading to a reduction in mobility at high operating temperatures.
- μ_{surf} reflects surface-roughness scattering, which occurs at high transverse electric fields (high V_{GS}) and is highly dependent on the quality of the SiC surface and oxidation processing.

The degradation of μ_{eff} has a profound impact on the operational characteristics and reliability of SiC MOSFETs:

Resistance Partitioning: Unlike Silicon devices where channel resistance (R_{ch}) is often negligible, the low mobility in SiC causes R_{ch} to account for a substantial portion (up to 30–50%) of the total specific on-resistance in 650 V and 1200 V devices.

Transconductance and Switching Speed: Mobility directly determines the transconductance (g_m). Low μ_{eff} results in a lower g_m, which can prolong the Miller plateau during switching, potentially increasing switching losses and affecting di/dt control.

Short-Circuit Robustness: Higher channel resistance inherently limits the peak saturation current during a fault. While this increases conduction losses, it provides a beneficial increase in the short-circuit withstand time (t_{SC}), allowing more time for protection circuits to react.

Temperature Coefficient: At high temperatures, the dominance of phonon scattering causes mobility to decrease. This results in a positive temperature coefficient for $R_{DS(on)}$, which is essential for stable current sharing when paralleling multiple MOSFETs in high-power modules.

Dynamic measurements under fast gate transitions confirm that mobility and threshold voltage are history-dependent, particularly after high-field stress or high-temperature operation, due to the slow dynamics of interface traps.

Table 2.2 summarizes the qualitative contribution and system-level impact of these mobility-limiting mechanisms.

Table 2.2: Qualitative contribution and impact of mobility-limiting mechanisms in SiC MOSFETs.

Mechanism	Strength	Primary Impact	Notes
Coulomb Scattering	High	V_{th} hysteresis & R_{ch}	Dominant at low–moderate fields
Phonon Scattering	Medium	$R_{DS(on)}$ Temp. Coeff.	Increases at high T
Surface Roughness	Med.–Low	Peak g_m limits	Process dependent

2.4.3 Subthreshold Slope and Gate-Leakage Scaling: Planar vs. Trench Architectures

The electrostatics and field distributions near the gate oxide differ significantly between planar and trench SiC MOSFETs. These differences directly influence the subthreshold behavior—the "turn-on" transition region—and the long-term reliability of the gate dielectric.

Subthreshold Slope and the Capacitive Tug-of-War The subthreshold slope (S) represents the gate voltage required to increase the drain current by one order of magnitude (measured in mV/decade). Physically, this value is determined by a competition between the gate's ability to control the surface potential and the "parasitic" capacitances that resist that control. In SiC, the slope is modeled as:

$$S \approx \underbrace{(\ln 10)\,\frac{kT}{q}}_{\text{Thermal Limit}} \times \underbrace{\left(1 + \frac{C_{\text{it}} + C_{\text{dep}}}{C_{\text{ox}}}\right)}_{\text{Capacitive Divider}} \tag{2.4}$$

The equation reveals two distinct components:

- **The Thermal Limit ($\approx$ 60 mV/dec):** This is the theoretical minimum at room temperature. Even in a "perfect" device with zero defects, the switch cannot turn on sharper than this limit.
- **The Capacitive Divider:** This term represents the "efficiency" of the gate. In an ideal device, this multiplier would be 1. However, C_{it} (interface-trap capacitance) and C_{dep} (depletion capacitance) act as a "tug-of-war" against the gate oxide capacitance (C_{ox}).

In Silicon Carbide, the interface trap density (D_{it}) is orders of magnitude higher than in Silicon. This results in a massive C_{it} value. Physically, as the gate voltage increases, a large portion of the charge is "wasted" filling up interface traps instead of creating the conductive inversion layer. This "stretches out" the turn-on phase, leading to a higher (worse) S value, increased threshold voltage (V_{th}) variability, and higher leakage in the "off" state.

Architectural Impact: Planar vs. Trench The geometric transition from planar to trench architectures changes how these traps interact with the device physics:

- **Crystal Plane Advantage:** Planar MOSFETs usually utilize the Si-face (0001) of the crystal. Trench MOSFETs form their channels along the vertical sidewalls, often utilizing the a-face $(11\bar{2}0)$ or m-face $(1\bar{1}00)$. These vertical planes often exhibit lower D_{it}, which can reduce C_{it} and provide a sharper subthreshold slope compared to planar designs.
- **Field Crowding and Leakage:** While the trench offers better slope characteristics, it introduces "corners" in the architecture. These corners cause electric field crowding. Without advanced shielding (such as the P+ shield regions shown in Figure 2.1), this high field can lead to "soft" subthreshold behavior, where gate leakage increases prematurely before the device is fully turned on.

Architectural Impact: Planar vs. Trench The transition from planar to trench architectures represents a fundamental shift in how the inherent anisotropy of the 4H-SiC crystal is utilized. 4H-SiC is a hexagonal crystal structure, and its electronic properties—specifically the density of interface traps—vary significantly depending on the crystal plane utilized for the channel.

- **Crystal Plane Anisotropy (The "Top" vs. "Side" Surface):** Planar MOSFETs are fabricated on the **Si-face** (the (0001) basal plane). This surface is characterized by a high density of dangling chemical bonds, which results in a high concentration of interface traps (D_{it}) when the gate oxide is grown. These traps serve as Coulombic scattering centers, significantly reducing the inversion-layer mobility.

  In contrast, a Trench MOSFET utilizes the **vertical sidewalls** (typically the a-face or m-face) for the channel. These vertical planes naturally possess a lower density of dangling bonds, resulting in a cleaner interface with reduced D_{it}. This material advantage allows for higher carrier mobility and a more decisive subthreshold transition.

- **Geometric Field Enhancement:** While the vertical architecture improves channel transport, the geometry introduces structural vulnerabilities. In a planar device, the gate oxide is a flat interface, which promotes a uniform electric field distribution. In a trench device, the sharp corners at the base of the trench act as sites for electric field crowding. These "corners" concentrate the field lines, subjecting the dielectric to higher localized stress than the nominal oxide field would suggest.

Quantifying Gate-Leakage Scaling As the localized electric field at these trench corners increases, the probability of carriers tunneling through the oxide barrier rises. This behavior is governed by the Fowler–Nordheim relationship:

$$J = A \cdot E_{\text{ox}}^2 \cdot \exp\left(-\frac{B}{E_{\text{ox}}}\right) \tag{2.5}$$

The parameters are defined as follows:

- A **(Pre-exponential constant):** A parameter determined by the effective mass of the tunneling carrier and the specific barrier height of the SiC/SiO_2 interface.

- B **(Fowler–Nordheim constant):** This represents the "critical field" for tunneling. It is physically defined by the barrier height (ϕ_b) and the effective mass (m^*) of the carrier. Because E_{ox} is in the denominator of a negative exponent, the leakage current (J) is exponentially sensitive to localized field increases.

Design implications. Tighter subthreshold control improves parallel-device predictability and light-load behavior, while gate leakage provides an early indicator of oxide degradation pathways developed in Chapter 3.

Design Trade-offs and Shielding This physical relationship establishes a critical engineering trade-off. While the trench architecture exploits the vertical crystal planes to solve the mobility and subthreshold slope issues inherent in planar Si-face designs, it simultaneously creates regions of high field intensity. Because of the exponential nature of Equation 2.5, even minor geometric non-uniformities at a trench corner can lead to orders-of-magnitude increases in leakage current and accelerated oxide wear-out. To mitigate this, high-performance trench designs must incorporate P+ shielding regions to "clamp" the electric field and protect the trench corners from the high drain potentials.

Electrostatic control and corner-field effects in trench gates. Trench MOSFETs introduce two electrostatic features that are largely absent in planar devices: (i) *corner fields* at trench bottoms and upper corners, and (ii) stronger coupling between the gate and the JFET/drift regions because the gate wraps around the current path. These effects can improve channel control (reducing the effective depletion contribution seen in S) while simultaneously increasing local oxide field. The practical outcome is that a trench device can exhibit a steeper apparent turn-on in transfer characteristics, yet it may also show earlier onset of field-driven leakage unless oxide thickness, nitridation, and corner rounding are optimized.

Role of near-interface traps and temperature scaling. In SiC, the energy distribution of interface states near the conduction-band edge often dominates subthreshold behavior. Because C_{it} is bias- and temperature-dependent, the measured slope $S(T)$ is not a single constant but a signature of trap density *and* trap time constants relative to measurement bandwidth. A robust comparison between planar and trench architectures therefore benefits from measuring S over temperature and frequency to separate quasi-static trap response from high-frequency response.

Gate leakage: Fowler–Nordheim vs. trap-assisted paths. Gate leakage may be governed by Fowler–Nordheim tunneling at high fields, but at moderate fields trap-assisted tunneling and defect-assisted conduction can dominate. Trench devices can experience higher local fields near corners, shifting the onset of leakage to lower applied V_{GS} if corners are sharp or if oxide quality is non-uniform. A practical comparison is therefore not only $\log(I_G)$ vs. V_{GS}, but $\log(I_G)$ vs. estimated oxide field and temperature.

Design implication. Subthreshold slope and gate leakage are coupled through the same interface quality and local-field distribution. Process steps that reduce D_{it} and improve electrostatics typically improve both S and long-term V_{th} stability, linking Chapter 2 electrostatics directly to Chapter 3 lifetime models.

2.4.4 Nitridation and Mobility Enhancement

Post-oxidation annealing in nitric oxide (NO) or N_2O atmospheres reduces interface-trap density by roughly an order of magnitude and can double or triple channel mobility, depending on orientation and process. More recent work explores alternative gate stacks (e.g., thin interfacial SiO_2 with high-k top layers) and plasma-based treatments to further suppress D_{it}, with mixed success in long-term reliability tests.

From a power-converter perspective, mobility improvements translate directly into lower $\mathrm{R}_{\mathrm{DS(on)}}$ at a given voltage rating, especially in the 650–900 V classes where the drift contribution is relatively small. However, these benefits must be balanced against potential impacts on oxide reliability, as aggressive treatments can modify defect populations relevant to TDDB and BTI.

2.4.5 JFET Region and Cell Geometry

For planar MOSFETs, the JFET region between adjacent p-wells contributes significantly to total resistance. Reducing JFET resistance requires:

- narrower cell pitch and optimized p-well spacing,
- deeper p-wells for better field control,
- carefully tuned channel and JFET doping profiles.

Trench MOSFETs eliminate most of the traditional JFET constriction by moving the channel vertically along the trench sidewalls, which increases effective channel perimeter and reduces channel and JFET components of $R_{DS(on)}$. However, these architectures introduce stronger electric-field crowding near trench corners, linking channel design more tightly to oxide reliability, as discussed later.

2.4.6 Temperature Dependence of Channel and Drift Resistance

SiC MOSFETs exhibit a characteristic temperature dependence in on-resistance:

- Drift-region resistance typically *decreases* modestly with temperature due to increased carrier concentration in lightly doped regions.
- Channel resistance and JFET resistance *increase* with temperature as mobility degrades, especially under strong Coulomb and phonon scattering.

Measurements and electrothermal simulations in multiple papers show that $R_{DS(on)}$ can increase by 1.5–2.5× between 25°C and 175°C, with the exact factor depending on device structure and manufacturer. High-fundamental-frequency motor drives and high-temperature modules (e.g., 200°C junction) must therefore account for this increase in conduction loss when defining derating strategies and mission-profile lifetime.

2.5 Switching Dynamics of SiC MOSFETs

SiC MOSFETs exhibit exceptionally fast switching transitions due to their majority-carrier conduction, low gate charge, and high saturation velocity. These advantages enable high-frequency conversion and reduced switching losses but introduce challenges in layout, packaging, and gate-drive design. Fast transitions produce high dv/dt (often exceeding 50 kV/μs) and high di/dt (over 5 A/ns), which in turn increase susceptibility to parasitic-induced voltage overshoot, current imbalance, and EMI.

A full understanding of switching behavior requires analyzing:

- the intrinsic MOSFET capacitances and their voltage dependence,
- gate-charge dynamics (Q_g partitioning),
- parasitic inductances in the power and gate loops,
- the role of common-source inductance (CSI),
- dv/dt-induced gate coupling and crosstalk,
- the interaction between body diode, MOS channel, and drift region.

These topics are treated in detail in the following subsections and closely connect to gate-drive strategies and converter-level EMI design, which are further expanded in other Chapters.

2.5.1 Intrinsic Capacitances

SiC MOSFETs possess highly non-linear internal capacitances that are fundamental to their high-speed switching performance. These parasitic components include:

- **Gate–Source Capacitance (C_{gs}):** Primarily determined by the overlap of the gate electrode over the source regions and the channel; it remains relatively constant across varying voltage levels.

- **Gate–Drain Capacitance (C_{gd}):** Also known as the *Miller capacitance*, this component is extremely non-linear. It decreases rapidly as the drain-source voltage (V_{DS}) increases and the depletion region expands into the JFET and drift regions.
- **Drain–Source Capacitance (C_{ds}):** Associated with the p-n junction between the body and the drift region.

In power electronics design, these are often grouped into composite terminal capacitances:

- **Input Capacitance ($C_{iss} = C_{gs} + C_{gd}$):** Dictates the charge required from the gate driver to transition the device.
- **Output Capacitance ($C_{oss} = C_{ds} + C_{gd}$):** Varies sharply with increasing voltage. While a low C_{oss} at high voltage is advantageous for resonant Zero Voltage Switching (ZVS) transitions, its extreme non-linearity complicates accurate switching loss modeling.
- **Reverse Transfer Capacitance ($C_{rss} = C_{gd}$):** Directly influences the feedback from the drain to the gate during switching.

Figure 1.2 presents a conceptual C–V curve reflecting this typical behavior. Because the Gate–Drain capacitance (C_{gd}) dominates at high voltages, it serves as the primary governor of switching characteristics:

- **Turn-off dynamics:** Typically faster than turn-on because the depletion region expands rapidly, causing C_{gd} to shrink and allowing the drain voltage to rise quickly.
- **Turn-on dynamics:** Heavily constrained by the *Miller plateau* (the period where gate current is diverted to charge C_{gd}) and layout parasitics. Overshoot and ringing in the gate and drain waveforms depend strongly on the power loop and gate loop inductances.
- **dv/dt sensitivity:** The voltage slew rates are extremely sensitive to the external gate resistance (R_g) and the specific gate-driver topology utilized.

These capacitive relationships differentiate SiC MOSFETs from Silicon IGBTs, where minority-carrier storage (tail current) plays the dom-

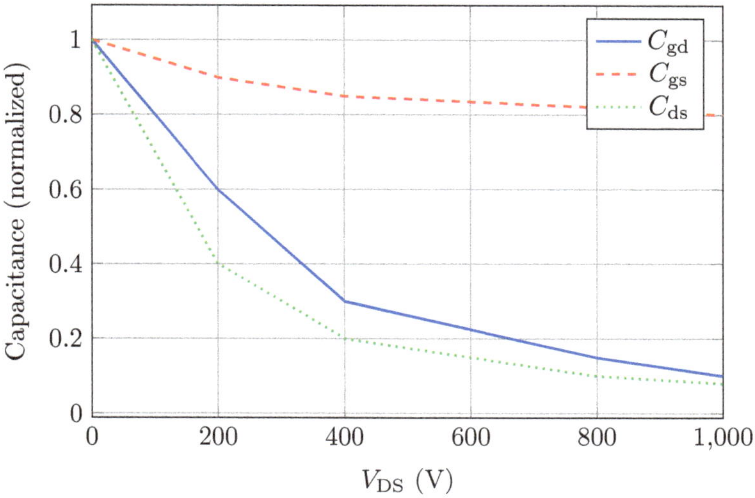

Figure 2.3: Illustrative nonlinear capacitance–voltage behavior of SiC MOSFETs. The rapid decrease of C_{gd} with voltage strongly influences switching speed.

inant role in switching dynamics, whereas SiC dynamics are almost entirely determined by the charging and discharging of these non-linear intrinsic capacitances.

2.5.2 Gate-Charge Dynamics and the Miller Plateau

The detailed definition and practical implications of the Miller plateau, including driver-design consequences, are consolidated in Chapter 5 (Control & Gate-Drive Master). This chapter retains the material-physics perspective and focuses on how capacitances and charge storage arise from device structure.

2.5.3 Power-Loop and Gate-Loop Parasitics

Definitions and layout-centric treatment of gate-loop inductance and its interaction with switching waveforms are consolidated in Chapter 5. Here we note only that parasitics couple device-level charge dynamics to system-level ringing and overvoltage.

2.5.4 Turn-On and Turn-Off Dynamics

Turn-on dynamics are governed by the interplay between gate-drive strength, drift-region depletion charge, Miller capacitance, and package inductances. A typical sequence during turn-on is:

1. Gate voltage rises toward threshold; the channel forms and current begins to rise.
2. Drain current ramps up toward load current while V_{DS} remains high.
3. The Miller plateau occurs as V_{DS} collapses and the drain current may overshoot due to parasitics.
4. Overshoot and ringing appear depending on loop inductance and stray capacitances.

Turn-on losses E_{on} typically dominate over turn-off losses, especially in topologies where the body diode or an antiparallel SiC diode conducts just before turn-on.

Turn-off behavior is generally faster than turn-on because the depletion region expands rapidly and C_{gd} decreases with increasing V_{DS}. However, turn-off can still exhibit substantial voltage overshoot and ringing if the load current and loop inductances are high. Detailed transient modeling methods, including SPICE and FPGA-based real-time models, are discussed in several papers that support converter-level design and hardware in the loop (HIL) testing.

2.5.5 dv/dt-Induced Gate Coupling and False Turn-On

Due to the high switching speeds of SiC MOSFETs, high voltage slew rates (dv/dt) can couple through the parasitic Gate–Drain capacitance (C_{gd}) and induce unintended charging of the gate-source terminal. This phenomenon is often approximated by the following relationship:

$$\Delta V_{GS} \approx C_{gd} \cdot \frac{dV_{DS}}{dt} \cdot Z_{gate} \tag{2.6}$$

where:

- ΔV_{GS} is the induced transient voltage at the gate-source terminal.
- C_{gd} is the highly non-linear Miller capacitance.

- dV_{DS}/dt is the voltage slew rate across the drain-source terminals.

- Z_{gate} is the total impedance of the gate loop, including external gate resistance and internal gate-driver impedance.

If this induced ΔV_{GS} exceeds the device threshold voltage (V_{th}), it leads to a false turn-on (or parasitic turn-on). This is a critical concern in half-bridge configurations—the standard building block of phase-legs—where the rapid turn-on of one device (the "aggressor") can cause a transient voltage spike on the gate of the complementary device (the "victim") that should remain off. If both devices conduct simultaneously, a shoot-through event occurs, leading to excessive losses or catastrophic failure.

Modern mitigation strategies for dv/dt-induced crosstalk include:

- **Kelvin-Source Connections:** A dedicated fourth pin that decouples the power-current loop from the gate-control loop. This prevents the high di/dt in the power circuit from inducing a voltage across the common-source inductance, which would otherwise counter the gate-driver signal.

- **Negative Turn-off Gate Voltages:** Driving the gate to a negative bias (e.g., -2 V to -5 V) during the off-state. This provides a larger safety margin between the induced ΔV_{GS} and the threshold voltage V_{th}.

- **Split-Output Gate Drivers:** Utilizing drivers with separate turn-on and turn-off paths (typically using an external diode) to allow for independent optimization of switching speeds and gate-loop damping.

- **Minimization of L_{gate}:** Reducing the gate-loop inductance (L_{gate}) and the physical area of the local gate loop to minimize the total impedance Z_{gate} and reduce susceptibility to electromagnetic interference (EMI).

These techniques are widely adopted in modern active gate-driver designs and high-density module pinouts to maintain signal integrity despite the extreme dv/dt levels associated with wide-bandgap (WBG) semiconductors.

2.5.6 Energy Loss Mechanisms and Planar vs. Trench Devices

Switching energy comprises:

$$E_{sw} = E_{on} + E_{off} + E_{rr}, \tag{2.7}$$

where:

- E_{on}: turn-on energy,
- E_{off}: turn-off energy,
- E_{rr}: diode reverse-recovery energy.

SiC Schottky diodes virtually eliminate E_{rr}, making switching efficiency primarily a function of parasitics and gate drive. When the intrinsic body diode is used, however, reverse recovery and bipolar degradation significantly increase both loss and reliability risk, as discussed in the next section.

Trench devices generally switch faster than planar designs due to reduced JFET resistance and higher transconductance, but they experience stronger electric-field crowding in the oxide. Table 2.3 summarizes qualitative switching differences. The increased switching performance of trench devices must be balanced against their higher oxide stress and potentially reduced avalanche capability.

Table 2.3: Switching performance comparison: Planar vs. Trench SiC MOSFETs.

Criteria	Planar	Trench	Technical Notes
Switching Speed	Moderate	Superior	Vertical crystal planes and cell density result in higher g_m.
Voltage Overshoot	Moderate	Elevated	High di/dt amplifies inductive spikes ($L \cdot di/dt$) across parasitics.
Gate-Oxide Stress	Lower	Higher	Field crowding at trench corners increases risk of oxide degradation.
dv/dt **Control**	Flexible	Challenging	High g_m increases susceptibility to parasitic oscillations.

2.6 Conclusion

Chapter 2 has detailed the physics of the 4H-SiC MOSFET, demonstrating how its wide bandgap and high critical field translate into superior power density and high-frequency capability. However, this same physics introduces unique reliability "fingerprints":

- **Interface Sensitivity:** The high density of interface traps (D_{it}) at the oxide layer directly dictates the Bias Temperature Instability (BTI) and threshold voltage shifts that will be analyzed in Chapter 3.
- **Bipolar Risk:** The intrinsic *p–i–n* body diode, while functional, carries the risk of stacking-fault expansion, which can permanently degrade the crystal structure under conduction stress.
- **Thermal Flux:** The high current densities enabled by SiC necessitate advanced thermal management to prevent the hotspot-driven short-circuit failures described in the next chapter.

By understanding these fundamental device behaviors, we can now move to Chapter 3, which focuses on long-term reliability engineering—bridging the gap between the semiconductor physics presented here and the long lifetime requirements of modern power electronic systems.

Chapter 3

Reliability of SiC MOSFETs

3.1 Introduction

This chapter is the *Reliability & Failure Master* for SiC MOSFETs. It connects the device physics introduced in Chapter 2 to measurable degradation modes, lifetime extrapolation, and safe operating area constraints. Where Chapter 2 focuses on intrinsic material and channel behavior, Chapter 3 quantifies how those properties degrade under electrical, thermal, and environmental stress. Practical protection implementations and gate-drive control techniques are treated in Chapter 5, and Chapter 6 applies these reliability and control constraints to full converter architectures.

3.2 Gate-Oxide Reliability in SiC MOSFETs

The integrity of the gate oxide is perhaps the most scrutinized aspect of SiC technology. Extensive research has addressed the physics of Time-Dependent Dielectric Breakdown (**TDDB**—the gradual wear-out of the oxide under constant field) and Bias Temperature Instability (**BTI**—the shift in threshold voltage over time due to trapped charge).

Unlike Silicon MOSFETs, which benefit from a nearly defect-free oxide interface, SiC oxides suffer from intrinsic material limitations:

- **High Interface Trap Density (D_{it}):** Densities in the range of 10^{12} to 10^{13} $\text{cm}^{-2}\text{eV}^{-1}$ lead to significant charge trapping.
- **Carbon-Related Defects:** Thermal oxidation often leaves be-

hind carbon interstitials (C_i), carbon substitutions (C_s), or carbon clusters, which act as "near-interface" traps.

- **Broad Time Constants:** Traps vary from "fast" (responding in microseconds) to "slow" (responding over hours), making threshold voltage (V_{th}) behavior highly history-dependent.
- **Thermally Activated Breakdown:** High operating temperatures increase the probability of electrons gaining enough energy to tunnel into or damage the oxide lattice.

These defects manifest as V_{th} hysteresis, mobility degradation, and stochastic (random) breakdown behavior. This is particularly critical in trench MOSFETs, where the gate oxide is exposed to complex multidimensional fields, and the oxide quality directly limits the maximum permissible gate-driver voltage.

3.2.1 Electric-Field Distribution: Planar vs. Trench MOSFETs

The reliability profile of a device is largely dictated by its geometry. Planar MOSFETs distribute the vertical electric field relatively uniformly across the gate oxide. This results in:

- **Predictable TDDB:** Failure follows a consistent statistical distribution with a high "Weibull slope" (meaning failures are tightly clustered in time, indicating a controlled wear-out process).
- **Standardized Stress Levels:** Compatibility with traditional 15–18 V gate-drive signals for a 20-year lifetime.

In contrast, Trench MOSFETs (including U-shaped or V-shaped "VMOSFET" variants) experience **Field Crowding**. The electric field lines concentrate at the bottom corners of the trench, creating localized oxide stress that can exceed the average design value by 30–60%. This leads to:

- **Localized Oxide Rupture:** Higher susceptibility to "soft" breakdown (increased leakage) or "hard" breakdown (catastrophic shorts).
- **Sensitivity to Parasitics:** Increased vulnerability to gate-voltage overshoots caused by common-source inductance and high switching speeds (dv/dt).

These geometric differences explain the divergent reliability strategies seen in commercial data sheets and are central to the design of advanced shielding structures utilized in modern SiC architectures.

3.3 Bias Temperature Instability (BTI)

Bias Temperature Instability (BTI) is a parametric degradation mechanism that manifests as a semi-permanent shift in the threshold voltage (V_{th}) when the gate is subjected to a constant DC bias at elevated temperatures. Unlike Silicon devices, where BTI is primarily a concern for P-channel MOSFETs, SiC N-channel MOSFETs exhibit significant shifts under both Positive Bias (PBTI) and Negative Bias (NBTI).

The root cause of BTI in the SiC/SiO_2 system is the interaction between mobile carriers in the channel and a diverse population of traps:

- **Interface Traps (D_{it}):** Located exactly at the SiC/SiO_2 boundary, these "fast traps" communicate with the channel on microsecond scales.
- **Border Traps (Near-Interface Oxide Traps):** Located within 1–3 nm of the oxide, these "slower" traps allow carriers to tunnel in and out, with time constants ranging from milliseconds to kiloseconds.
- **Fixed Oxide Charges:** Deeper defects that lead to a permanent, irreversible shift in the device's baseline V_{th}.

During operational stress, PBTI (positive gate bias) causes electrons to be injected from the channel into the oxide, increasing V_{th} and subsequently increasing the on-resistance ($R_{DS(on)}$). Conversely, NBTI (negative gate bias) can lead to a V_{th} decrease, which poses a significant reliability risk as it reduces the device's noise immunity and increases the risk of parasitic turn-on.

3.3.1 Time Dependence and Modeling

The evolution of the threshold-voltage shift (ΔV_{th}) in SiC MOSFETs does not reach a steady state quickly; rather, it follows a characteristic logarithmic trend over several decades of time. This behavior is typically modeled using the empirical relationship:

$$\Delta V_{th}(t) = A \cdot \ln\left(1 + \frac{t}{t_0}\right) \tag{3.1}$$

where:

- A is the magnitude coefficient, which is a function of the oxide electric field, the junction temperature, and the specific density of defects in the gate stack.
- t is the cumulative stress time.
- t_0 is a time constant representing the transition between the filling of "fast" interface traps and "slow" border traps.

This logarithmic nature (illustrated in Figure 3.1) implies that the most aggressive shifts occur during the first few minutes of operation. For engineers, this makes "burn-in" testing and early-life screening critical to ensure that V_{th} stabilizes within a window compatible with the gate-driver's logic levels.

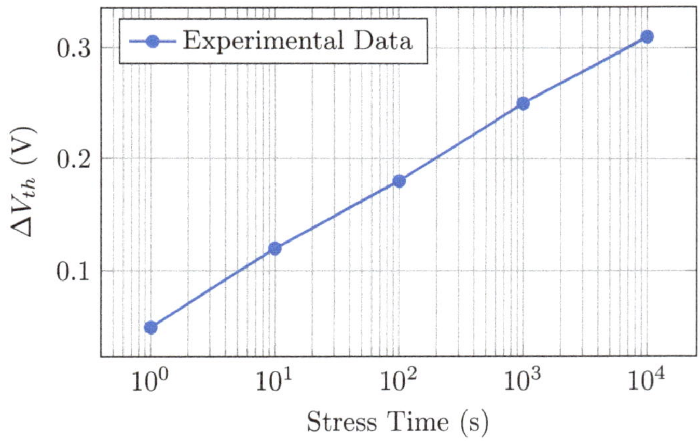

Figure 3.1: Typical threshold-voltage shift under PBTI stress. The trend highlights the rapid early-stage trapping followed by long-term logarithmic growth.

3.3.2 Acceleration Factors: Field and Temperature

The rate of BTI degradation is highly sensitive to the operating environment. Accelerated Life Testing (ALT) utilizes extreme conditions to extrapolate the long lifetime of a device:

- **Field Acceleration:** Increasing the gate bias (e.g., from +18 V to +30 V) heightens the oxide electric field, lowering the energy barrier for carrier tunneling into border traps.
- **Thermal Acceleration:** Elevated temperatures (125°C to 175°C) provide the thermal energy required for carriers to overcome deep trap barriers, significantly increasing the coefficient A in Equation 3.1.

3.3.3 Recovery Behavior and Hysteresis

A unique and challenging aspect of BTI in Silicon Carbide is its *recovery behavior*. Once the stress voltage is removed, the trapped charge begins to emit back into the channel, causing V_{th} to "relax" toward its original value.

This recovery is also logarithmic and is categorized into two components:

1. **Recoverable Component:** Charges trapped in interface and near-border traps that discharge within seconds or minutes. This creates a "hysteresis" effect during high-frequency switching.
2. **Permanent Component:** Charges trapped deep within the oxide or structural damage to the SiC/SiO_2 lattice that does not recover, representing the true "aging" of the device.

In practical applications, recovery is heavily influenced by the mission profile. For instance, the transition from body-diode conduction to the off-state can alter the local field distribution, effectively "resetting" or accelerating the recovery of V_{th}. To obtain robust design margins for automotive traction inverters, composite models—combining reaction-diffusion theory with charge-trapping dynamics—are utilized to predict V_{th} stability over millions of switching cycles.

3.4 Time-Dependent Dielectric Breakdown (TDDB)

Time-Dependent Dielectric Breakdown (TDDB) is a fundamental wear-out mechanism where the gate oxide (SiO_2) undergoes a gradual loss of insulating integrity under long-term electric field stress, eventually resulting in a catastrophic short circuit. In Silicon Carbide (SiC) MOSFETs, TDDB is a critical reliability benchmark because these devices operate at internal electric fields significantly higher than those found in traditional Silicon counterparts.

The heightened sensitivity of SiC oxides to TDDB is driven by four primary factors:

- **High Nominal Fields:** To achieve competitive $R_{DS(on)}$, SiC MOSFETs utilize thinner gate oxides relative to the high blocking voltages they handle.
- **Geometric Field Enhancement:** Sharp geometries, particularly in trench MOSFETs, act as "lightning rods," concentrating the electric field.
- **Charge Injection:** Higher interface trap densities (D_{it}) facilitate the injection of carriers into the oxide lattice, accelerating the formation of a conductive "percolation path."
- **Thermal Acceleration:** High-temperature operation provides the energy necessary for ionic and electronic defects to migrate and coalesce into a failure path.

3.4.1 Intrinsic vs. Extrinsic Breakdown Modes

Reliability engineers distinguish between failure regimes based on whether the breakdown is a result of fundamental physics or manufacturing defects. Table 3.1 provides a side-by-side comparison of these two modes, highlighting their divergent origins and statistical distributions.

1. **Intrinsic Breakdown:** This represents the fundamental limit of the SiO_2 material. It is the result of the cumulative breaking of Silicon-Oxygen ($Si - O$) bonds due to high-energy carrier bombardment. This mode is deterministic and follows a tight statistical distribution.
2. **Extrinsic Breakdown:** This mode is caused by "infant mortality" defects introduced during fabrication, such as residual carbon clusters or plasma-etch damage. These defects create local weak spots where the oxide is effectively thinner, leading to premature failure.

As shown in Table 3.1, the statistical spread (β) is the primary indicator used during production to verify if a batch of devices is limited by process defects (Extrinsic) or by the inherent properties of the oxide (Intrinsic).

Table 3.1: Comparative characteristics of breakdown modes in SiC MOSFETs.

Feature	Intrinsic Breakdown	Extrinsic Breakdown
Physical Cause	Lattice wear-out	Material/Process defects
Predictability	High (Deterministic)	Low (Stochastic)
Lifecycle Phase	End-of-life (Wear-out)	Early-life (Infant mortality)
Statistical Spread	Narrow ($\beta > 20$)	Wide ($\beta < 10$)

3.4.2 Weibull Statistics and Manufacturing Consistency

The cumulative breakdown probability $F(t)$ is modeled using the Weibull distribution, which allows engineers to predict failure rates over a 20-year lifespan:

$$F(t) = 1 - e^{-(t/\eta)^{\beta}} \tag{3.2}$$

where:

- t is the time under stress.
- η **(Characteristic Life):** The time at which 63.2% of the population has failed.
- β **(Weibull Slope):** A shape parameter indicating the "tightness" of the failure distribution.

A high β value indicates a mature manufacturing process. In SiC, Planar MOSFETs typically exhibit $\beta \approx$ 20–35 due to their uniform fields, while Trench MOSFETs often show lower values ($\beta \approx$ 5–15) due to the variability introduced by trench etching and corner field crowding.

3.4.3 Field-Acceleration Models for Lifetime Extrapolation

To predict a 20-year lifetime from a 1,000-hour test, high-field data must be extrapolated using physical acceleration models.

The E-Model (Linear Field Acceleration)

The "E-model" is the most widely adopted model for planar SiC MOSFETs. It is based on the thermochemical model of bond breaking, where the electric field lowers the activation energy required to break the $Si - O$ bonds. The relationship is spelled out as:

$$\ln(TTF) = A - \gamma E_{ox} \tag{3.3}$$

where:

- TTF **(Time-to-Failure):** The predicted time until the oxide ruptures (usually the median life).
- A**:** A constant that incorporates the junction temperature (T_j) and the inherent quality of the oxide material.
- E_{ox}**:** The electric field across the gate oxide (typically measured in MV/cm).
- γ **(Field Acceleration Parameter):** This is the "slope" of the reliability curve. It represents the sensitivity of the oxide lifetime to changes in voltage; a larger γ means that even a small increase in gate voltage will lead to a massive reduction in operational life.

The Power-Law Model

While the E-model is standard for planar devices, the "Power-Law model" is frequently utilized for Trench MOSFETs:

$$TTF = C \cdot (E_{ox})^{-n} \tag{3.4}$$

In this model, the exponent n accounts for the fact that local field enhancements at trench corners do not scale linearly with the applied gate voltage. This model often provides a better fit for complex geometries where multi-dimensional fields dominate the stress profile.

By applying these models, manufacturers can define the "Safe Operating Area (SOA)" for the gate voltage, ensuring that the probability of failure remains below the "six-sigma" thresholds required for automotive and aerospace applications.

3.4.4 Planar vs. Trench TDDB Behavior

The architectural differences between planar and trench SiC MOSFETs lead to divergent Reliability-of-Failure (RoF) profiles. While both utilize the same gate-oxide material (SiO_2), the geometric orientation of the electric field relative to the SiC crystal lattice is fundamentally different.

Planar MOSFETs are characterized by a "mature" reliability profile. Because the gate oxide is grown on the flat (0001) Si-face of the crystal, they exhibit:

- **Uniform Field Distribution:** Field lines remain largely perpendicular to the interface, resulting in a predictable and manageable stress profile across the entire gate area.
- **Predictable Scaling:** The Time-Dependent Dielectric Breakdown (TDDB) follows standard thermochemical scaling over temperature and field, allowing for high-confidence lifetime projections.
- **Process Maturity:** Modern planar fabrication has minimized extrinsic defects, such as interface roughness or carbon inclusions, leading to higher effective oxide lifetimes at a given field.

Trench MOSFETs are utilized to maximize cell density and eliminate JFET resistance, but they introduce complex electrostatic challenges:

- **Localized Stress Points:** The vertical orientation means that field lines "crowd" at the trench corners and bottom. Without mitigation, these points experience fields 30–50% higher than the nominal value.
- **Geometric Variability:** The "Weibull slope" (β) is typically lower in trench devices ($\beta \approx 5$–15), indicating a wider statistical spread in failure times due to microscopic variations in trench etching and sidewall roughness.
- **Advanced Shielding Requirements:** To preserve the oxide, manufacturers employ "deep P-shields" or "floating P-wells." These structures are designed to "clamp" the electric field and shield the

trench bottom from the high drain-source voltage during the off-state.

3.4.5 Soft vs. Hard Breakdown in SiC MOSFETs

TDDB failure in Silicon Carbide does not always manifest as an instantaneous, catastrophic rupture. Instead, it is categorized into two distinct physical regimes based on the current-conduction dynamics:

Soft Breakdown (SBD)

Soft breakdown is an incremental failure where localized conduction filaments form within the oxide, but they do not yet possess enough energy to cause physical melting.

- **Consequence:** A gradual increase in gate-leakage current (I_{GSS}).
- **System Impact:** While the device may remain "operational," the V_{th} often shifts, and the *noise margin* is severely degraded. This makes the device more susceptible to "parasitic turn-on" from Miller capacitance transients.

Hard Breakdown (HBD)

Hard breakdown is the final, catastrophic stage of failure where a low-impedance path is permanently established.

- **Consequence:** Immediate oxide rupture, leading to a massive surge in current between the gate and the drain or source.
- **System Impact:** This is almost always destructive. In a bridge topology, an HBD failure in the gate can bypass protection and destroy the gate driver, potentially leading to a DC bus short-circuit fault.

3.4.6 Lifetime Extrapolation and Safe Operating Area (SOA)

To establish the gate-voltage Safe Operating Area (SOA), manufacturers utilize Time-Dependent Dielectric Breakdown (TDDB) qualification based on Weibull statistics and field-acceleration models. These models, primarily the E-model and Power-Law model, enable the extrapolation of the Time-to-Failure (TTF) from short-term, high-stress conditions to the intended operational life of the system.

As illustrated in Figure 3.2, the relationship between the applied oxide field (E_{ox}) and the expected lifetime is exponential. The intersection of this acceleration curve with the required life defines the maximum permissible continuous gate-source voltage (V_{GS}) for a given temperature.

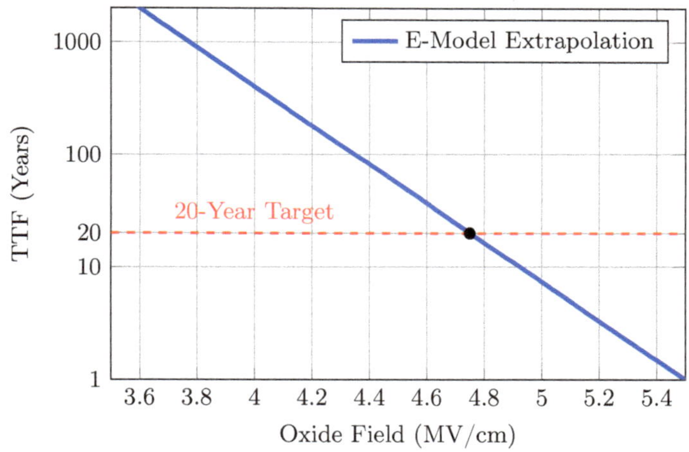

Figure 3.2: Lifetime extrapolation for SiC MOSFET gate-oxide reliability. The intersection between the field-acceleration curve and the required target (e.g., 20 years) determines the maximum reliable operating field and corresponding V_{GS}.

Current industry benchmarks for 1200 V-class MOSFETs indicate a divergence in reliability strategies based on device architecture:

- **Planar Architectures:** These devices typically demonstrate a TTF exceeding 20 years at a nominal +18 V bias, even at junction temperatures (T_j) of 175°C. The inherent stability of the planar Si-face interface makes these devices a preferred choice for industrial and grid-scale applications where long-term reliability is prioritized over die-size reduction.

- **Trench Architectures:** While offering superior on-resistance, trench designs often target a 10–15 year lifetime at the same +18 V nominal bias due to localized field enhancement. To meet the stringent lifetime requirements of automotive standards (e.g.,

AEC-Q101), designers frequently implement gate-voltage derating, limiting the maximum operational V_{GS} to +15 V.

These characteristics underscore that the performance advantages of Silicon Carbide are intrinsically linked to a "reliability budget." Successful implementation requires a multi-disciplinary approach that balances precise gate-drive control with robust thermal management to maintain the oxide field within these safety limits.

3.5 Body-Diode Degradation and Bipolar Reliability

Although the SiC MOSFET is primarily a unipolar device, its structural architecture inherently contains a parasitic p–n junction formed between the p-well and the n-type drift region. When this "body diode" is forward-biased—common during dead-time in bridge topologies—it behaves as a bipolar p–i–n diode. This operation mode triggers a unique reliability concern known as **bipolar degradation**, a phenomenon largely absent in Silicon MOSFETs but critical in the SiC material system.

Forward conduction in the body diode introduces three primary failure-relevant stressors:

1. **Minority-Carrier Injection:** The injection of holes into the n-drift region facilitates the activation and expansion of crystal defects.

2. **Recombination-Enhanced Dislocation Motion (REDM):** Energy released during electron-hole recombination acts as a catalyst for structural changes in the SiC lattice.

3. **Reverse-Recovery Dynamics:** Unlike SiC Schottky diodes, the body diode must remove stored minority carriers during turn-off, resulting in a reverse-recovery charge (Q_{rr}) that increases switching losses and generates inductive voltage overshoots.

3.5.1 Mechanism of Bipolar Degradation: BPD to SSF Transition

Bipolar degradation is a structural reliability phenomenon unique to wide-bandgap semiconductors like SiC. It is characterized by the irreversible expansion of crystalline defects that increase the device's electrical resistance. The process is driven by the transition of **Basal-Plane Dislocations (BPDs)** into **Shockley Stacking Faults (SSFs)**.

The "Seed" and the "Fuel"

The root of the problem lies in **Basal-Plane Dislocations (BPDs)**—small, line-like misalignments in the SiC crystal lattice created during the ingot growth or epitaxial process. Under normal "unipolar" operation (where only electrons carry current), these BPDs remain stationary and dormant.

However, when the MOSFET's internal body diode is activated, the device enters "bipolar" conduction, where both electrons and holes are injected into the drift region. When these carriers meet and recombine, they release energy. In SiC, this recombination energy is sufficient to act as a mechanical "fuel," enabling the BPDs to glide through the crystal lattice.

Physical Progression of Degradation

The transformation occurs through a process known as *recombination-enhanced dislocation glide*:

1. **Carrier Recombination:** During body-diode conduction, electron-hole pairs recombine in the N-drift region.

2. **Dislocation Glide:** The energy released from recombination allows a BPD to split into two "partial" dislocations. One of these partials glides across the basal plane of the crystal.

3. **Stacking Fault Formation:** As the dislocation moves, it leaves behind a **Shockley Stacking Fault (SSF)**—a planar defect where the atomic stacking sequence of the SiC is out of order (e.g., ABCB instead of ABAC).

4. **Current Path Obstruction:** These SSFs act as high-resistivity barriers. Because the SSF has a locally wider bandgap than the surrounding 4H-SiC, it creates a potential barrier that repels carriers. This "pinches" the available area for current flow, effectively reducing the conductive cross-section of the device.

Electrical and System-Level Consequences The physical expansion of SSFs manifests as a "parametric drift" in the device's performance:

- **Resistance Increase:** A permanent increase in the on-state resistance ($R_{DS(on)}$), which leads to higher conduction losses and increased thermal stress.
- **Diode Instability:** An increase in the forward voltage drop (V_f) of the body diode.

While modern "BPD-free" epitaxial growth techniques have significantly reduced the density of these starting "seeds," bipolar degradation remains a critical design constraint. It is particularly relevant for applications involving heavy inductive loads and frequent freewheeling diode conduction, such as traction inverters and motor drives, where the cumulative effect of diode usage can lead to premature system failure.

3.5.2 Reverse-Recovery Stress and Switching Overload

When the body diode transitions from conduction to a blocking state (reverse recovery), the minority carriers must either recombine or be swept out by the electric field. This process is quantified by the reverse-recovery charge (Q_{rr}), the peak reverse current (I_{rrm}), and the reverse-recovery time (t_{rr}).

In SiC MOSFETs, the body-diode Q_{rr} is significantly higher than that of a dedicated SiC Junction Barrier Schottky (JBS) diode. Repeated exposure to these high-energy turn-off events leads to:

- **Localized Self-Heating:** High instantaneous power dissipation during t_{rr} occurs in the p-well/drift region junction.
- **Secondary Stress:** The resulting voltage overshoots ($V_{pk} = L \cdot di/dt$) stress the gate-oxide and can trigger the BTI mechanisms discussed in Section 3.3.
- **Coupled Degradation:** There is strong evidence that high-frequency reverse-recovery stress accelerates the BPD-to-SSF transition, linking switching dynamics directly to crystal stability.

3.5.3 Mitigation Strategies

To circumvent the reliability risks associated with bipolar conduction, designers typically adopt one of three strategies:

- **Synchronous Rectification:** Actively turning on the MOSFET channel during the freewheeling period. Since the channel is unipolar, current flows through the majority-carrier path, bypassing the p–n junction and preventing hole injection.
- **External JBS Diodes:** Placing a SiC Schottky diode in anti-parallel with the MOSFET. The lower turn-on voltage ($V_{f\_Schottky} \approx$ 1.2V) compared to the body diode ($V_{f\_body} \approx$ 3.0V) ensures the body diode never enters forward conduction.
- **Dead-Time Optimization:** Minimizing the duration of body-diode conduction through precise gate-timing control, reducing the total "charge budget" injected into the drift layer over the device's lifetime.

3.5.4 Bipolar Degradation and Stacking-Fault Expansion

While the previous sections outlined the electrical inefficiency of the body diode, its most significant long-term impact is a material reliability issue known as bipolar degradation. Widely documented in early generations of 4H-SiC unipolar devices, this failure mode remains an essential consideration for high-reliability mission profiles.

The degradation mechanism is driven by the recombination energy of minority carriers, which facilitates the motion of crystal defects. This process follows a specific physical sequence:

1. **Activation:** Minority-carrier injection during body-diode conduction or avalanche events provides the energy required to activate dormant basal-plane dislocations (BPDs) within the epitaxy.
2. **Transformation:** These BPDs dissociate into partial dislocations, creating single Shockley stacking faults (SSFs) that reside on the basal plane of the 4H-SiC crystal.
3. **Lateral Expansion:** Under continued conduction, the SSFs expand laterally across the active area. Because the stacking fault

has a different electronic structure than the host crystal, it acts as a local quantum-well-like barrier.

4. **Local Conductivity Modulation:** The faulted regions exhibit a higher local leakage current and a lower turn-on voltage, which redistributes the current density unevenly across the die.

5. **Permanent Degradation:** The cumulative effect is a permanent, irreversible increase in $R_{DS(on)}$ and a shift in the forward voltage (V_F) of the body diode, eventually leading to thermal runaway or device failure.

Table 3.2 summarizes the experimental trends and the resulting system-level impacts of this degradation.

Table 3.2: Key behaviors associated with bipolar degradation in 4H–SiC MOSFETs.

Phenomenon	Device-Level Impact	Root Mechanism
Stacking-fault growth	Permanent $R_{DS(on)}$ increase	BPD Glide/Activation
Increased Forward Drop	Higher conduction losses	p–i–n recombination
Carrier Storage	Elevated Q_{rr} and E_{rr}	Bipolar injection
Avalanche Wear-out	Reduced UIS ruggedness	High-field carrier stress

To combat these effects, modern industrial practices employ a combination of material-level and system-level mitigation strategies:

- **Material Optimization:** Substantial reduction of BPD density in the starting substrate and the subsequent epitaxial growth. Advanced "BPD-free" epitaxy techniques have become a standard for automotive-grade components.

- **System-Level Control:** Implementing synchronous rectification to carry reverse current through the majority-carrier MOS channel, thereby preventing the body diode from entering bipolar conduction.

- **Carrier Lifetime Engineering:** Utilizing proton irradiation or "lifetime killers" to reduce the recombination lifetime of minority carriers, effectively starving the stacking faults of the energy needed for expansion.

- **External Integration:** Bypassing the intrinsic diode entirely with an external anti-parallel SiC Schottky Barrier Diode (SBD), especially in mission profiles involving heavy freewheeling.

While modern 1200 V automotive-grade SiC MOSFETs have largely mitigated catastrophic bipolar degradation through superior material processing, conservative design remains necessary. In mission-critical sectors such as traction and aerospace, limiting body-diode stress is still a primary design rule to ensure a long operational lifetime.

3.6 Avalanche Ruggedness and UIS Testing

The Unclamped Inductive Switching (UIS) test is the industry-standard benchmark for evaluating a MOSFET's "avalanche" capability—its ability to survive overvoltage transients that exceed its rated breakdown voltage. Unlike a clamped circuit where a diode or MOV absorbs energy, in a UIS event, the energy stored in the circuit's stray or load inductance is forced directly into the semiconductor.

As illustrated in the Figure 3.3, the UIS process follows a precise sequence:

1. **Energy Accumulation:** While the MOSFET is on, current builds up linearly in the series inductor (L).

2. **Interruption:** The MOSFET is turned off. Because the inductor current cannot drop to zero instantaneously, the drain-source voltage (V_{DS}) rises rapidly.

3. **Avalanche Clamping:** V_{DS} reaches the device's actual breakdown voltage ($V_{(BR)DSS}$). At this point, the SiC lattice undergoes "impact ionization," where high-energy carriers collide with atoms to create electron-hole pairs, allowing current to flow even though the channel is off.

4. **Energy Dissipation:** The inductive energy is converted into heat within the device's drift region and body diode.

SiC MOSFETs generally demonstrate superior avalanche ruggedness compared to Silicon (Si) counterparts. This is primarily due to

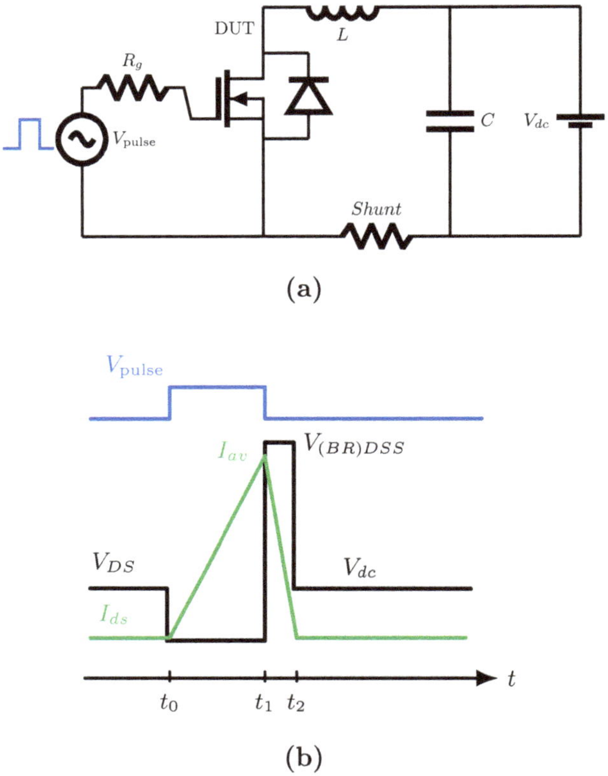

Figure 3.3: Avalanche test configuration and corresponding idealized switching waveforms for a SiC MOSFET. (a) Double-pulse test circuit. (b) Waveforms showing avalanche current (I_{av}) during the turn-off transient.

SiC's higher critical electric field (which allows for higher breakdown voltages) and its wide bandgap, which enables the rapid removal of generated carriers without the prolonged "minority-carrier tail currents" that cause excessive heating in Si devices.

However, architecture plays a major role: trench devices often experience "field crowding" at the gate corners. This localized stress can lead to premature failure compared to planar designs, effectively lowering the "energy budget" the device can handle before catastrophic failure.

3.6.1 Avalanche Current Paths and Hotspot Locations

During an avalanche event, current is not distributed uniformly. It flows through the body diode and drift region, creating localized thermal hotspots. The failure modes associated with these hotspots include:

- **Gate Oxide Rupture:** Localized heating at the interface can weaken the SiO_2 layer, leading to a gate-to-source short.
- **Parasitic BJT Latch-up:** Extreme heat can activate the parasitic Bipolar Junction Transistor (BJT) inherent in the MOSFET structure. Once the BJT turns on, the gate loses the ability to turn the device off, leading to immediate destruction.
- **Drift-Layer Filamentation:** Current can "crowd" into a narrow filament, leading to localized melting.

As a result, device-layout parameters—such as cell pitch and JFET region geometry—must be optimized to spread this energy as evenly as possible across the die surface.

3.6.2 Avalanche Energy and Safe Operating Area

The amount of energy a device can absorb during a single pulse is termed the "Single-Pulse Avalanche Energy (E_{AS})". It is calculated as:

$$E_{AS} = \frac{1}{2} L I_{peak}^2 \left(\frac{V_{(BR)DSS}}{V_{(BR)DSS} - V_{dc}} \right),$$

where I_{peak} is the peak current at the start of the avalanche, $V_{(BR)DSS}$ is the breakdown voltage, and V_{dc} is the DC supply voltage. In many simplified analyses, the voltage-clamping ratio is omitted, and the energy is approximated as $\frac{1}{2} L I_{peak}^2$.

Based on experimental data, E_{AS} is highly sensitive to the operating environment:

- **Temperature Sensitivity:** As the initial junction temperature (T_j) increases, E_{AS} decreases because the device reaches its critical melting or latch-up temperature more quickly.
- **Architectural Variance:** Trench MOSFETs typically exhibit 20–40% lower E_{AS} than planar devices of the same rating due to the corner field stresses mentioned earlier.

3.7 Terrestrial Neutron-Induced Single Event Burnout (SEB) and Altitude Effects

Terrestrial neutrons (produced by cosmic-ray interactions in the atmosphere) can initiate localized ionization tracks and trigger impact-ionization driven failure in high-voltage blocking states, particularly under high electric field.

3.7.1 Mechanism

A neutron-induced secondary particle can deposit charge in the drift region and locally enhance carrier density. Under sufficiently high field, this can initiate a regenerative avalanche condition. The resulting localized heating and current filamentation may drive catastrophic burnout if the event exceeds transient thermal containment capability.

3.7.2 Altitude dependence

Neutron flux increases with altitude and geomagnetic latitude. Applications such as aerospace power converters and high-altitude operation therefore experience elevated exposure relative to sea level. Mission profiles should include environment-dependent SEB risk for high-voltage blocking periods.

3.7.3 Mitigations

Mitigation typically combines electrical margining (avoid high-field excursions), overshoot control, screening, and application-specific qualification for environments with elevated neutron flux.

3.7.4 Atmospheric neutron spectrum and altitude scaling

Terrestrial neutrons arise from cosmic-ray interactions in the atmosphere. Neutron flux increases strongly with altitude because atmospheric shielding is reduced, and the energy spectrum shifts toward higher energies. In practical terms, an operating point that is safe at sea level can exhibit a non-negligible SEB rate in high-altitude aircraft, mountainous regions, or aerospace environments.

3.7.5 Device-level SEB mechanism in SiC MOSFETs

SEB is initiated when a high-energy neutron produces a nuclear reaction that generates energetic recoils and a dense ionization track. Under high V_{DS}, the induced carriers can trigger a localized conductive filament and a rapid current transient. The resulting electrothermal feedback produces a hotspot that can exceed material limits on a sub-millisecond timescale. In SiC MOSFETs, sensitivity is often linked to high-field regions in the drift/JFET area and to transient overshoot, making layout and switching-shape control indirectly relevant to SEB risk.

3.7.6 Engineering model and practical design margin

A pragmatic engineering model treats SEB event rate as proportional to neutron flux and an effective device cross-section that depends on blocking voltage, field distribution, and temperature. System-level mitigation therefore focuses on (i) voltage derating, (ii) limiting transient overshoot, and (iii) ensuring uniform current distribution during high-field stress. For altitude exposure, derating envelopes can be scaled by flux factors and validated with conservative margin.

3.7.7 Screening and qualification considerations

Because SEB is a low-probability, high-consequence event, qualification evidence typically combines accelerated exposure (beam tests) with mission-profile extrapolation. Applications with significant altitude exposure should document the assumed neutron environment and include SEB in the safety case alongside short-circuit SOA and avalanche limits.

3.8 Short-Circuit Reliability and SOA Constraints

Short-circuit (SC) events are arguably the most violent electrical stresses a SiC MOSFET faces. While SiC's material properties are superior, its physical implementation creates a "thermal mass" challenge. Because SiC dies are significantly smaller than equivalent Silicon dies, they have a much lower thermal capacitance. This means that during a fault, the junction temperature (T_j) rises at a staggering rate of 100–200°C/µs.

The "Short-Circuit Withstand Time (t_{SC})"—the duration a device can survive a dead-short before failing—is consequently much shorter than for Si IGBTs. Table 3.3 compares the short circuit performance of Planer and Trench technologies in SiC MOSFETs.

Table 3.3: Comparative short-circuit (SC) performance of 1200 V planar and trench SiC MOSFETs.

Parameter	Planar (1200 V)	Trench (1200 V)
Typical withstand time, t_{SC}	5–8 μs	2–4 μs
Peak sat. current, I_{sat}	Mod. (lower g_m)	High (higher g_m)
Primary failure mechanism	Gate-oxide rupture	Thermal runaway / Al melt
Required protection target	$< 5\,\mu$s	$< 2\,\mu$s

To ensure system reliability, these times must be strictly greater than the sum of the gate-driver's fault-detection time and its reaction time (blanking time + shutdown time).

3.8.1 Electrothermal Evolution and Failure Mechanisms of Short-Circuit Events

The physical progression of a short-circuit event in a SiC MOSFET is a high-speed electrothermal process where the device must dissipate a massive energy density within a few microseconds. This evolution is categorized into four distinct stages driven by the feedback between junction temperature (T_j) and carrier transport.

Stage 1: Initial Peak and Saturation Upon fault inception, the drain current (I_D) surges rapidly toward its saturation level. This peak value is determined by the gate-source voltage (V_{GS}) and the device transconductance (g_m). Due to their higher channel density, trench-gate architectures typically exhibit significantly higher initial current peaks than planar variants. At this stage, the current is limited by "velocity saturation," where the extremely high electric field causes carriers to reach their maximum scattering-limited velocity, regardless of further increases in drain voltage.

Stage 2: Thermal Self-Limiting (Current Droop) As the junction temperature surges—often exceeding 500°C within the first 2–3 μs the carrier mobility (μ) decreases significantly due to increased lattice scattering. In SiC, mobility typically scales with temperature as $\mu \propto T^{-2.5}$. This mobility degradation increases the channel resistance, resulting in a characteristic "droop" in the current waveform. This phase provides a critical, albeit brief, safety window where the device partially self-regulates its power dissipation.

Stage 3: Thermal Runaway Onset If the fault is not cleared within the withstand time (t_{SC}), the localized temperature may reach a point where the intrinsic carrier concentration (n_i) approaches the background doping level (N_D). At this threshold, the semiconductor begins to lose its doped characteristics and behaves like an intrinsic conductor. The gate lose control over the channel, and the current begins to rise uncontrollably as the device enters a positive feedback loop of heating and carrier generation.

Stage 4: Catastrophic Failure Mechanisms Final failure typically manifests through one of three primary physical pathways:

- **Gate-Oxide Rupture:** The extreme thermo-mechanical stress caused by the rapidly expanding SiC lattice, combined with high localized electric fields, leads to the dielectric breakdown of the gate oxide.

- **Parasitic BJT Latch-up:** The high current and temperature can trigger the parasitic NPN bipolar junction transistor (BJT) inherent in the MOSFET structure. Once "latched," the current can no longer be interrupted by the gate, leading to a permanent short-circuit.

- **Metallization Melting:** In high-energy events, the aluminum source metallization may reach its melting point (660°C). The molten metal can then penetrate the dielectric layers or the SiC surface, resulting in a catastrophic terminal-to-terminal short.

3.8.2 Gate-Drive Strategies for Short-Circuit Protection and Energy Management

Because the short-circuit withstand time (t_{SC}) of SiC MOSFETs is inherently limited by their small die area, the gate-driver serves as the primary safeguard for the power module. System reliability is determined by the driver's ability to manage the total energy (E_{SC}) dissipated during a fault before it reaches a critical threshold.

The total energy absorbed by the semiconductor during a short-circuit event is the time-integral of the instantaneous power dissipation:

$$E_{SC} = \int_0^{t_{SC}} V_{DS}(t) \cdot I_D(t)\, dt \tag{3.5}$$

For a 1200 V device operating on a typical 800 V DC bus, E_{SC} can reach values between 1–4 J. Since the drain current in saturation (I_{sat}) is a direct function of the gate-source voltage (V_{GS}), the choice of gate bias significantly impacts ruggedness. A higher V_{GS} increases I_{sat}, which accelerates the energy accumulation rate and exponentially shortens the available t_{SC}.

To ensure the device remains within its Safe Operating Area (SOA), modern gate-drive architectures implement the following "active" protection strategies:

- **Desaturation (Desat) Detection:** This is the standard method for fault identification. The circuit monitors V_{DS} while the device is commanded "ON." If V_{DS} exceeds a specific threshold (indicating the device has left the ohmic region due to a fault), the driver triggers a shutdown sequence after a predefined "blanking time" intended to filter out switching noise.

- **Soft Turn-Off and Two-Level Turn-Off (2LTO):** During an SC event, the drain current is several times higher than the nominal rating. A rapid, standard turn-off would result in an extreme di/dt, generating a massive voltage spike ($V_{spike} = L_\sigma \cdot di/dt$) across the stray loop inductance (L_σ). This spike can easily exceed the breakdown voltage ($V_{(BR)DSS}$). To prevent this, "Soft Turn-Off" reduces V_{GS} slowly or in discrete steps (Two-Level Turn-Off), gradually decreasing the current to manage the inductive overshoot safely.

- **Active Current Limiting:** Some advanced drivers dynamically reduce the gate voltage immediately upon fault detection. By lowering V_{GS} to a level just above the threshold voltage (V_{th}), the driver "clamps" the short-circuit current to a manageable level, effectively slowing the rate of temperature rise and extending the t_{SC} budget.

By shifting the burden of reliability from the semiconductor lattice to the intelligent gate-driver circuitry, these active strategies can extend the effective short-circuit withstand capability of a system by 20–40%. This enables the use of high-performance trench MOSFETs in mission-critical applications where the native t_{SC} would otherwise be insufficient for standard protection hardware.

3.8.3 Hotspot Formation

Failure is rarely uniform across the die. Localized "hotspots" form due to microscopic non-uniformities in cell density or threshold voltage (V_{th}). Because V_{th} typically has a negative temperature coefficient, a warmer cell will turn on more "hard," drawing more current and becoming even hotter. This positive feedback loop is a primary concern for trench MOSFETs, where the high cell density and vertical geometry can concentrate electric fields at the trench corners, accelerating the onset of thermal runaway.

3.9 Electrothermal Instability and Thermal Runaway

Although Silicon Carbide (SiC) is recognized for its superior thermal conductivity and wide-bandgap resilience, SiC MOSFETs remain susceptible to electrothermal instability. This phenomenon occurs when the localized heat generated within the semiconductor lattice exceeds the heat-extraction capacity of the device packaging and cooling system, establishing a divergent positive feedback loop known as thermal runaway.

The onset of electrothermal instability is governed by the competition between two temperature-dependent parameters: the threshold voltage (V_{th}) and the carrier mobility (μ). In SiC MOSFETs, V_{th} typically possesses a negative temperature coefficient, meaning it decreases as the temperature rises, which tends to increase the drain current. Conversely, carrier mobility decreases with temperature due to increased phonon scattering, which acts to reduce the current.

Thermal runaway is triggered when the following two criteria are simultaneously met:

1. **Positive Temperature Coefficient of Current:**

$$\frac{\partial I_D}{\partial T} > 0 \tag{3.6}$$

   At high current densities or specific gate-bias conditions, the current increase caused by the V_{th} shift outweighs the current reduction from mobility degradation. This causes the device to "draw" more current as it heats up.

2. **Thermal Impedance Imbalance:**

$$P_{\text{gen}} > \frac{T_j - T_a}{R_{th(j-a)}} \tag{3.7}$$

   The instantaneous power dissipation ($P_{\text{gen}} = V_{DS} \cdot I_D$) exceeds the rate of heat removal through the thermal resistance network (R_{th}) to the ambient environment (T_a).

Electrothermal coupling is particularly acute during high-energy transient events, such as short circuits or Unclamped Inductive Switching (UIS). In these scenarios, the junction temperature (T_j) can experience excursions of several hundred degrees Celsius within microseconds. Because this heating is effectively adiabatic on such short timescales, the cooling system—including the Thermal Interface Material (TIM) and heatsink—cannot respond fast enough to mitigate the rise.

The risk is further amplified in trench-gate architectures. The significantly higher cell density and narrower pitch of trench designs concentrate the power dissipation into a smaller volume compared to planar devices. This concentration creates localized "hotspots" that can trigger filamentation—a state where current crowds into a small, overheated region of the die—leading to localized lattice melting or gate-dielectric failure before the average die temperature reaches critical limits.

3.10 Thermo-Mechanical Reliability and Power Cycling

The reliability of a SiC system is not solely determined by the semiconductor die; the surrounding package often fails first due to thermo-mechanical fatigue. Power cycling—the repeated heating and cooling of the device during operation—induces mechanical stress due to the Coefficient of Thermal Expansion (CTE) mismatch between the SiC die, the copper lead frame, and the ceramic substrate.

3.10.1 Die-Attach Reliability: Solder vs. Sintered Silver

The die-attach layer is the primary thermal conduit between the SiC MOSFET die and the module's ceramic substrate or heat sink. While standard Silicon devices often rely on traditional solder, the extreme power densities and elevated junction temperatures characteristic of Silicon Carbide demand a more robust interface. Consequently, high-performance SiC modules have largely transitioned from solder alloys to **Sintered Silver (Ag)** technology.

Thermal and Mechanical Superiority: Sintered silver has become the industry standard for SiC packaging due to two key physical advantages:

- **Conductivity:** With a thermal conductivity exceeding 200 W/m·K, sintered Ag is approximately four times more efficient at heat extraction than standard SnAgCu (SAC) solders.
- **Homologous Temperature:** Silver's melting point is significantly higher (962°C) than that of SAC solder ($\approx$ 217°C). In a system operating at 175°C, solder is physically "near" its melting point, making it highly susceptible to creep and grain coarsening. Sintered silver operates at a much lower fraction of its melting point (homologous temperature), ensuring long-term structural stability.

Modeling Thermomechanical Fatigue: Fatigue in the die-attach layer stems from the Coefficient of Thermal Expansion (CTE) mis-

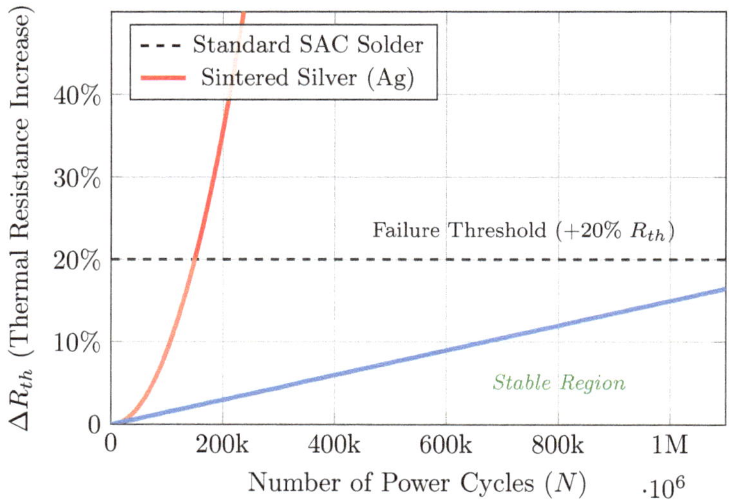

Figure 3.4: Comparative thermal degradation of die-attach materials. Solder typically fails within 10^5 to 2×10^5 cycles for $\Delta T_j = 100\,\text{K}$, while sintered silver exhibits superior thermomechanical stability beyond 10^6 cycles.

match between the SiC die and the substrate. This creates cyclic plastic strain ($\Delta\epsilon$) during every power cycle. To predict the operational life (N_f) of the interconnect, the **Coffin–Manson relationship** is utilized:

$$N_f = C \cdot (\Delta\epsilon)^{-c} \tag{3.8}$$

The physical significance of the parameters is defined as follows:

- N_f**:** The expected number of cycles until failure (typically defined by a 20% rise in R_{th}).
- C**:** A material constant influenced by the die-attach geometry and ductility.
- $\Delta\epsilon$**:** The range of plastic strain induced per cycle, dictated by the temperature excursion (ΔT_j).
- c**:** The fatigue exponent, which characterizes the material's sensitivity to large temperature swings.

As shown in Figure 3.4, sintered silver maintains a stable thermal resistance for far longer than solder. Its superior grain structure reduces the accumulation of plastic strain per cycle, extending the module's reliable service life by roughly an order of magnitude.

3.10.2 Interconnect Reliability: Bond-Wires vs. Clip-Bonds

In traditional power modules, heavy aluminum (Al) bond wires have served as the standard electrical interconnect for the source and gate terminals. However, the unique operational environment of Silicon Carbide (SiC)—characterized by extreme power densities and high switching speeds (dv/dt)—exposes the inherent mechanical and electrical limitations of wire-bonding.

Failure Mechanisms in Wire Bonds: The primary reliability concerns for bond wires in SiC applications are **heel cracking** and **bond-wire lift-off**. These failures are driven by the Coefficient of Thermal Expansion (CTE) mismatch between the aluminum wire (23×10^{-6}/K), the SiC die (4×10^{-6}/K), and the encapsulation material.

- **Heel Cracking:** Occurs at the junction where the wire transitions from the bond pad to the span. Repetitive thermal expansion causes the wire to flex, leading to fatigue and eventually a transverse crack.
- **Bond-Wire Lift-off:** Shear stresses at the interface between the Al wire and the SiC metallization eventually cause the bond to peel away, increasing the contact resistance and accelerating thermal failure.

The Shift to Copper Clip-Bonding: To meet the rigorous standards of automotive traction inverters, many manufacturers are transitioning to copper (Cu) clip-bond or planar interconnect technologies. Unlike wires, which provide point-contacts, clips utilize a larger surface area to distribute current and heat more uniformly across the die.

As summarized in Table 3.4, the transition to clip-bonding addresses two critical SiC challenges:

1. **Electromagnetic Performance:** The reduction in parasitic inductance from ~10 nH to less than 2 nH is vital. In high-speed SiC switching, even small inductances can trigger severe voltage ringing and electromagnetic interference (EMI).

Table 3.4: Comparative reliability and performance of interconnect technologies.

Metric	Al Bond Wire	Cu Clip-Bond	Advantage
Cycling Lifetime	Moderate	Very High	Clip reduces thermomechanical strain
Parasitic Inductance	High ($\sim$10 nH)	Low ($<$2 nH)	Essential for fast SiC switching
Thermal Distribution	Poor (top-side)	Excellent	Enables dual-side cooling paths

2. **Advanced Cooling:** Clips enable **Dual-Side Cooling (DSC)**. While bond wires only allow heat to escape through the bottom substrate, a copper clip acts as a top-side heat spreader, significantly lowering the overall junction-to-case thermal resistance ($R_{\theta JC}$).

3.11 Substrate Reliability: DBC and AMB Ceramics

The ceramic substrate provides essential electrical isolation while serving as the primary conduit for heat transfer from the SiC die to the baseplate. In Silicon Carbide applications, the increased junction temperatures ($T_j > 175°C$) and resulting thermal excursions place immense thermomechanical stress on the substrate, specifically at the interface between the ceramic and the copper metallization.

Bonding Technologies, DBC vs. AMB: Two primary technologies are utilized to bond the high-conductivity copper layers to the ceramic core:

- **Direct Bonded Copper (DBC):** This process relies on a eutectic melt of copper and oxygen (Cu–O) at approximately 1065°C. While cost-effective and standard for Alumina (Al_2O_3), DBC layers are prone to delamination and "cratering" (ceramic cracking) under the severe thermal cycling typical of SiC traction inverters.
- **Active Metal Brazing (AMB):** For Silicon Nitride (Si_3N_4), which does not naturally bond via the DBC process, AMB is

utilized. An active brazing filler creates a robust chemical bond that offers significantly higher mechanical integrity, making it the preferred choice for high-reliability SiC modules.

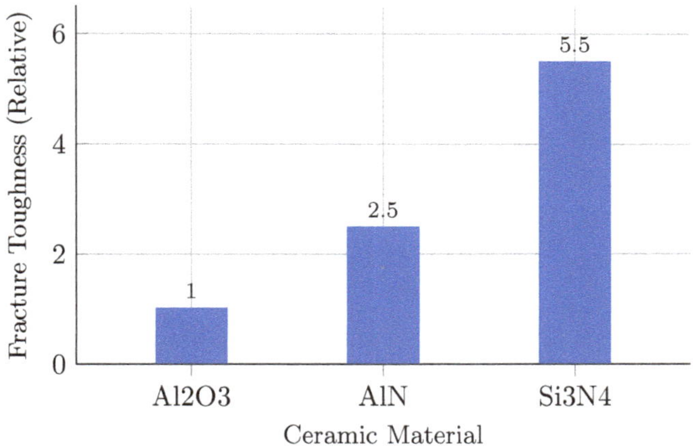

Figure 3.5: Relative reliability and mechanical strength comparison of ceramic substrates. Si_3N_4 offers a fracture toughness nearly five times that of Alumina, mitigating the "ceramic cratering" induced by high-power SiC switching.

Ceramic Material Performance: The choice of ceramic material dictates the trade-off between thermal resistance (R_{th}) and fracture toughness. While Aluminum Nitride (AlN) offers the highest thermal conductivity ($\sim 180\,\text{W/m·K}$), its low fracture toughness makes it brittle. Conversely, Silicon Nitride has emerged as the premier choice for automotive SiC applications due to its superior mechanical resilience.

The mismatch in Coefficient of Thermal Expansion (CTE) between the copper ($17\,\text{ppm/°K}$) and the ceramic (e.g., $3\,\text{ppm/°K}$ for Si_3N_4) is the primary driver of substrate wear-out. During power cycling, the superior fracture toughness of Si_3N_4 (as shown in Figure 3.5) allows it to withstand these shear stresses without the crack propagation that often leads to catastrophic isolation failure in Al_2O_3 DBC substrates.

3.12 Mission-Profile Modeling and Cumulative Damage

Mission-profile modeling serves as the "Digital Twin" bridge between real-world field data and the physics of failure. By translating electrical and thermal stress histories into lifetime consumption, designers can move beyond conservative "worst-case" assumptions toward optimized, usage-based reliability.

3.12.1 Thermal Fatigue and Lifetime Models

The primary failure mechanisms in SiC power modules under fluctuating loads are thermo-mechanical fatigue at the die-attach (sinter/solder) and the wire-bond interfaces. Because Silicon Carbide allows for higher power densities, the resulting temperature excursions (ΔT_j) are often more severe than in Silicon equivalents. The most robust model for capturing this fatigue is the **modified Coffin–Manson relationship**:

$$N_f = \alpha \cdot (\Delta T_j)^{-n} \cdot \exp\left(\frac{E_a}{k \cdot T_{j,m}}\right) \tag{3.9}$$

where the variables reflect the interaction between mechanical strain and chemical acceleration:

- N_f: Cycles until failure (typically a 20% rise in R_{th} or a 5% rise in $V_{CE,sat}$).
- α: A technology-specific constant (significantly higher for sintered Ag than for SAC solder).
- ΔT_j: The temperature swing ($T_{j,max} - T_{j,min}$). Due to the inverse-power relationship (exponent $n \approx 2$–5), doubling the ΔT_j can reduce the lifetime by a factor of 4 to 32.
- $T_{j,m}$: The mean junction temperature in Kelvin. This Arrhenius-term accounts for the fact that a 50°C swing is more destructive at a 100°C baseline than at 50°C due to increased atomic mobility and creep.
- E_a: Activation energy (eV), representing the energy barrier for the specific failure mode (e.g., crack propagation).

3.12.2 Application Case Studies: Stress Signatures

The "stress signature" of a mission profile dictates which failure model takes priority. Table 3.5 contrasts three distinct operational environments.

Table 3.5: SiC mission profiles and dominant reliability stressors.

Application	Typical Scenario	Cycle Char.	Reliability Priority
EV Traction	WLTP drive cycle	Rapid, deep ΔT_j	Die attach / interconnect fatigue
Solar/PV	Cloud shading	Periodic thermal transients	Gate-oxide wear-out (TDDB)
Data Center	Constant high load	Steady-state at high T_j	TIM degradation / aging

In an **EV Traction** inverter, the MOSFET experiences "active cycles" during every acceleration and "passive cycles" during ambient temperature changes. The high dv/dt of SiC can also introduce secondary stressors like dielectric aging. In contrast, a **Solar Inverter** faces "diurnal cycles" (once per day); here, the primary concern is not fatigue but the **Arrhenius-driven aging** of the gate-oxide and the drying of Thermal Interface Materials (TIM).

3.12.3 Thermal Aging: Arrhenius Acceleration

For non-cyclic, steady-state degradation, we utilize the Arrhenius Acceleration Factor (AF):

$$AF = \exp\left[\frac{E_a}{k}\left(\frac{1}{T_{use}} - \frac{1}{T_{test}}\right)\right] \tag{3.10}$$

This allows lifetime data from Accelerated Life Testing (ALT) at T_{test} (e.g., 200°C) to be scaled down to the actual operating temperature T_{use} (e.g., 125°C). This is vital for predicting the 20-year service life required for grid-tied infrastructure.

3.12.4 Cumulative Damage and Miner's Rule

To aggregate these diverse stresses into a single reliability metric, a five-step workflow is typically employed:

1. **Profile Conversion:** Convert mission data (speed/torque/irradiance) into a power dissipation profile $P_{loss}(t)$.
2. **Thermal Mapping:** Use a Cauer or Foster RC thermal network to generate the junction temperature history $T_j(t)$.
3. **Cycle Counting:** Apply the **Rainflow Counting Algorithm** to the $T_j(t)$ waveform. This reduces a chaotic temperature graph into a histogram of discrete cycles, each with a specific ΔT_j and $T_{j,m}$.
4. **Life Mapping:** For each bin in the histogram, calculate the cycles-to-failure N_i using Equation 3.9.
5. **Damage Accumulation:** Sum the partial damage using **Miner's Rule**:

$$D_{total} = \sum_i \frac{n_i}{N_i} \tag{3.11}$$

where n_i is the number of cycles actually experienced at stress level i. Failure is predicted when the cumulative damage D_{total} reaches unity (100%). In practice, automotive designers often target $D_{total} < 0.1$ (a $10\times$ safety margin) to account for material variances across a fleet.

3.13 Conclusion

This chapter has provided a comprehensive survey of the reliability landscape for Silicon Carbide MOSFETs. By examining the device from the atomic-level gate oxide to the macro-scale power module, several key conclusions emerge:

- **Gate Integrity:** The SiO_2 interface remains the "weakest link." Mechanisms such as BTI and TDDB require strict gate-voltage derating to ensure long lifetimes.
- **Bipolar Stability:** Body-diode conduction must be managed to prevent the irreversible expansion of stacking faults, which degrade the $R_{DS(on)}$ of the device.

- **Structural Ruggedness:** Trench architectures offer superior performance but require advanced shielding to match the avalanche and short-circuit ruggedness of planar designs.
- **Packaging as a Bottleneck:** High-temperature operation shifts the reliability burden to the package. Technologies like **sintered silver die-attach** and Si_3N_4 **substrates** are essential to prevent premature thermo-mechanical failure.

Ultimately, SiC reliability is not a static value but a "design budget." By understanding these physics-of-failure mechanisms, engineers can implement the advanced gate-drive and thermal management strategies detailed in the following chapters to fully unlock the potential of SiC technology.

Chapter 4

SiC Packaging Technologies and Thermal Management

4.1 Introduction

The packaging of Silicon Carbide (SiC) power semiconductor devices has become one of the most technically demanding and strategically important domains in power electronics design. Although enormous effort has been devoted to improving SiC MOSFET device structures, oxide reliability, channel mobility, and avalanche robustness, it is now widely recognized that the *practical* limits of SiC systems are overwhelmingly dictated not by the semiconductor die, but by the packaging technologies that surround it. Across the broader literature—including reliability studies, automotive teardown analyses, thermal modeling work, and packaging-focused device qualification reports—the consensus is clear: for SiC-based converters, packaging determines usable switching speed, operating temperature limits, long-term reliability, and system cost in almost every major application.

SiC MOSFETs are fundamentally capable of fast switching transitions (50–150 V/ns), high junction temperatures (175–225°C), and high power densities due to their small die size and low conduction losses. However, these same characteristics impose severe stresses on bond wires, die-attach layers, ceramic substrates, metallization interfaces, thermal interfaces, and cooling structures. Thermal gradients inside SiC modules can exceed 80–150 K during acceleration cycles in electric vehicles; dv/dt stress can exceed several tens of kilovolts per microsecond in aerospace converters; and power density in traction inverters has risen to 150–300 W/cm^2, forcing every thermal and mechanical layer to operate near its physical limits. As a result, packaging is increasingly the bottleneck in SiC module deployment rather than device physics itself.

The objective of this chapter is to develop a comprehensive and deeply technical treatment of packaging technologies and thermal management for SiC power modules.

4.2 Functional Role of Packaging in SiC Modules

The packaging in a SiC power module serves four tightly coupled functional roles: electrical performance, thermal management, mechanical reliability, and environmental protection. Unlike traditional silicon IGBT modules, where switching speeds, junction temperatures, and thermal gradients were modest, SiC's fast-switching and high-temperature capability push packaging into a multi-physics regime where electrical, thermal, and mechanical effects interact strongly.

4.2.1 Electrical Function: Parasitics and dv/dt Stress

SiC MOSFETs can switch with edge rates exceeding 50–150 V/ns under typical gate-drive conditions. Such high dv/dt produces displacement currents through stray capacitances, causing voltage overshoot, ringing, and EMI. Parasitic inductances as small as a few nH in the power loop can generate tens of volts of overshoot during turn-off transitions. Minimizing loop inductance is therefore essential, which directly links interconnect design, substrate layout, and module geometry to electrical performance.

Furthermore, high dv/dt stresses ceramic insulation layers, encapsulation materials, and inter-chip spacing. Partial-discharge (PD) inception voltages become critical, particularly in aerospace and high-altitude applications where insulation withstand strength decreases with pressure. Therefore, packaging must provide both low inductance and high insulation robustness.

4.2.2 Thermal Function: Heat Dissipation and Management

Silicon Carbide MOSFETs present a thermal paradox: while they operate with higher efficiency and lower absolute power losses than Silicon IGBTs, their significantly reduced die size results in much higher localized heat flux. For a typical 1.2 kV SiC die with an area of only a few tens of mm^2, the heat flux during peak transients (such as heavy EV acceleration) can exceed 200–300 W/cm^2. This intensity is nearly double that of traditional Silicon devices, making the thermal extraction path the primary limiting factor for power density.

The Thermal Extraction Chain: To prevent the junction temperature (T_j) from exceeding critical limits, heat must be moved through a multi-layered stack. Each interface represents a thermal bottleneck:

$$\text{Die} \rightarrow \text{Die Attach} \rightarrow \text{Substrate} \rightarrow \text{Baseplate} \rightarrow \text{TIM} \rightarrow \text{Cold Plate}$$

Thermal Resistance and Capacitance: The stack is typically modeled as an RC network (e.g., Cauer or Foster models).

- **Thermal Resistance (R_{th}):** Represents the steady-state temperature gradient across a layer. High-conductivity materials like sintered silver and Silicon Nitride (Si_3N_4) are used to minimize this value.
- **Thermal Capacitance (C_{th}):** Governs the transient thermal response. This is critical during short-circuit events or rapid torque pulses where the die generates heat faster than it can be conducted away, requiring the surrounding materials to "buffer" the energy.

Degradation and Thermal Runaway: The reliability of this pathway is interdependent. For instance, if the Die Attach begins to delaminate due to fatigue, the local R_{th} increases. This causes a localized "hot spot" on the die, which accelerates the chemical aging of the encapsulant and the thermal cycling stress on the substrate. Consequently, a failure in the micro-scale (die-attach) inevitably triggers a failure in the macro-scale (substrate or TIM).

4.2.3 Mechanical Function: Fatigue Resistance Under Cycling

Thermo-mechanical fatigue is the dominant module-level failure mechanism in SiC power modules. Large and rapid temperature swings cause:

- shear strain in the die attach,
- bending and tensile stress in the ceramic substrate,
- fatigue at the metal–ceramic interface,
- crack initiation and propagation in solder joints,
- bond-wire heel cracking (if wirebonds are used),
- void formation or pump-out in the TIM layer.

These stresses accumulate across thousands to millions of cycles, depending on the mission profile, making mechanical reliability one of the primary design challenges.

4.2.4 Environmental Protection

SiC modules deployed in EVs, wind turbines, photovoltaic inverters, or aerospace systems must survive exposure to humidity, condensation, salt spray, mechanical shock, and vibration. Encapsulation and conformal coatings must prevent tracking under high dv/dt, mitigate partial discharge, and provide stable dielectric properties across a wide temperature range.

4.3 Multi-Physics Finite Element Analysis (FEA) for Power Modules

Multi-physics Finite Element Analysis (FEA) provides a deterministic framework for predicting stress accumulation and identifying critical failure points in SiC modules. While analytical models provide a "first-pass" estimate, FEA is essential for capturing the spatial non-uniformity of heat flux and the complex thermomechanical interactions at the die-attach and substrate interfaces.

4.3.1 Recommended Simulation Hierarchy

A robust reliability assessment requires a sequential coupling of thermal and structural solvers. The recommended workflow is:

1. **Electro-Thermal Simulation:** Distributed losses (conduction and switching) are applied as heat sources to the die volume. This step computes the temperature field $T(\mathbf{r}, t)$, accounting for the temperature-dependent thermal conductivity of the materials.

2. **Thermomechanical Coupling:** The temperature field is mapped as a load into a structural solver. This simulation accounts for the Coefficient of Thermal Expansion (CTE) mismatch between the SiC die, the die-attach, and the ceramic substrate.

3. **Extraction of Damage Indicators:** Instead of looking at simple stress, engineers extract inelastic strain range ($\Delta\epsilon_{in}$) or strain-energy density (ΔW) per cycle. These metrics are then correlated with experimental power-cycling data to predict service life.

4.3.2 Simulating Die-Attach and Interconnect Stress

In SiC modules, the die-attach layer experiences the most significant stress due to the high dT/dt during switching. Simulation of these layers often requires non-linear constitutive models, such as "Anand's Model", to capture the rate-dependent creep and plastic flow of sintered silver or solder.

For modules utilizing "Dual-Sided Cooling (DSC)", the FEA boundary conditions must be adjusted to include thermal paths on both the top (clip) and bottom (substrate) sides. This configuration significantly reduces the peak T_j, but it introduces a "sandwich" stress state where the die is compressed between two rigid ceramic plates, potentially shifting the failure mode from bond-wire lift-off to die cracking.

4.3.3 Calibration and Validation

An FEA model is only as reliable as its input parameters. Validation is typically performed in two stages:

- **Thermal Calibration:** The simulated thermal impedance (Z_{th}) is compared against measurements from an Infrared (IR) camera or Transient Dual Interface (TDI) measurements. Discrepancies often point to poorly modeled contact resistances at the TIM interface.

- **Mechanical Correlation:** Predicted damage indicators are compared against post-mortem analysis of power-cycled modules. Techniques such as "Scanning Acoustic Microscopy (SAM)" are used to measure the actual delamination area, which is then used to tune the fatigue constants in the Coffin-Manson or Morrow models.

4.3.4 Integration with Mission-Profile Framework

The FEA results provide the "transfer function" between a mission profile and physical damage. By simulating a range of ΔT_j and $T_{j,m}$ values, a response surface can be generated. This surface allows for the real-time calculation of cumulative damage (Miner's Rule) using only the measured temperature telemetry from the inverter, effectively creating a health-monitoring system for the vehicle or power plant.

4.3.5 Solder Die Attach: Limitations and Fatigue Mechanisms

Solder alloys, typically Lead-free Sn-Ag-Cu (SAC) or high-lead compositions, have been the industry standard for decades in Silicon-based power modules due to their mature processing and low cost. However, the high power density of SiC MOSFETs subjects the die-attach to thermal conditions that push these materials to their physical limits.

Metallurgical Degradation Modes: Solder is a "low-melting-point" material relative to the 175°C–200°C operation of SiC. At these temperatures, solder operates at a high **homologous temperature** ($T_H = T_{op}/T_{melt} > 0.8$), making it highly susceptible to:

- **Inelastic Deformation:** Under cyclic thermal expansion, the solder undergoes both plastic and creep deformation as it attempts to accommodate the CTE mismatch between the SiC die (~ 4 ppm/K) and the substrate (~ 3–7 ppm/K).

- **Grain Coarsening and Softening:** Prolonged exposure to high T_j causes the microstructure to reorganize; small grains coalesce into larger ones, reducing the material's yield strength and facilitating crack path formation.

- **Intermetallic Compound (IMC) Growth:** Excessive heating accelerates the chemical reaction at the copper interface, forming brittle IMC layers (e.g., Cu_6Sn_5). While a thin IMC is necessary for bonding, a thick layer is brittle and acts as a site for crack initiation.

Fatigue Life Modeling: The accumulation of damage in the solder layer is primarily cycle-dependent. The fatigue life (N_f) is typically modeled using the **Coffin–Manson relationship**, which relates the number of cycles to failure to the plastic strain range ($\Delta\varepsilon_p$):

$$N_f = C \cdot (\Delta\varepsilon_p)^{-c} \tag{4.1}$$

In this context, c is the fatigue exponent (typically ≈ 2 for SAC alloys). In SiC applications where ΔT_j frequently exceeds 100 K, the resulting $\Delta\varepsilon_p$ is significantly higher than in traditional Si-IGBT modules. Consequently, while a Silicon module might reach 10^6 cycles, a soldered SiC module often reaches end-of-life at 10^3–10^5 cycles, failing to meet the reliability requirements of the automotive sector.

Table 4.1: Physical failure signatures in solder die-attach layers.

Failure Mode	**Primary Driver**	**Impact on SiC Performance**
Void Growth	Volumetric flow	Increased R_{th}, localized hot spots
Crack Prop.	Corner shear strain	Thermal runaway, V_{GS} instability
IMC Thickening	Diffusion (Arrhenius)	Brittle fracture at substrate interface
Ratchet Def.	Cyclic accumulation	Die tilt, warpage, delamination

As summarized in Table 4.1, the degradation of solder is an accelerating process. A small crack reduces the heat dissipation area, which increases the junction temperature, thereby further accelerating crack propagation—a positive feedback loop that ultimately leads to module failure.

4.3.6 Sintered Silver (Ag) Die-Attach

Sintered silver (Ag) has emerged as the premier die-attach technology for high-performance SiC power modules, effectively bridging the gap between the SiC die's high-temperature capabilities and the package's thermal limits. Unlike solder, which relies on a liquid-phase melt, silver sintering is a solid-state diffusion process where micron- or nano-scale Ag particles coalesce under heat and pressure to form a robust metallic joint.

Material Characteristics: The resulting sintered joint is a porous metallic matrix that provides a unique combination of electrical, thermal, and mechanical properties, as summarized in Table 4.2:

- **High Thermal Conductivity:** At 200–300 W/m·K, it provides a thermal path nearly an order of magnitude more efficient than SAC solder (~30–50 W/m·K).
- **Superior Thermomechanical Stability:** Because the melting point of the bulk silver joint remains at 962°C, the interconnect does not suffer from the creep and grain coarsening that plague solders at SiC operating temperatures (175°C–200°C).
- **Engineered Compliance:** The residual porosity (typically 5–30%) acts as a mechanical buffer, allowing the joint to absorb the shear strain induced by the CTE mismatch between the SiC die and the substrate.

Enabling SiC Performance: The adoption of sintered silver is not merely a reliability upgrade but a functional requirement for Silicon Carbide. The high power density of SiC MOSFETs generates localized heat flux that traditional solders cannot dissipate without exceeding safe junction temperatures (T_j). By significantly lowering the thermal resistance (R_{th}), sintered Ag allows the device to operate at the higher current densities enabled by SiC's superior Baliga Figure of Merit.

Furthermore, the high-temperature stability of the Ag-sinter joint ensures that the interconnect is no longer the "weakest link" in the system. This allows designers to fully exploit SiC's wide bandgap properties, supporting stable blocking voltages and low leakage currents at elevated temperatures without risking the mechanical integrity of the die-attach layer.

Table 4.2: Representative properties of sintered silver (Ag) die-attach materials (porosity-dependent).

Property	Typical Range	Design Relevance
Thermal cond. k	200–300 W/m·K	Crucial for managing extreme SiC heat flux
Elastic modulus E	20–60 GPa	Lower than bulk Ag; reduces stress on die
CTE	18–22 ppm/K	High mismatch requires careful substrate choice
Porosity	5–30%	Tunes mechanical compliance and fatigue life
Max continuous T	200–250°C	Enables operation beyond solder limits

4.4 Ceramic Substrates and Metallization Structures

The ceramic substrate in a SiC power module serves as the critical interface providing electrical isolation from the heat sink while acting as the primary thermal conduit for the die. In Silicon Carbide applications, the substrate must endure higher junction temperatures (T_j), steeper thermal gradients, and significantly higher heat fluxes compared to traditional Silicon IGBT modules. Consequently, the choice of both the ceramic core and the metallization process—Direct Bonded Copper (DBC) or Active Metal Brazing (AMB)—is a primary determinant of the module's mission-profile capability.

Metallization Architecture: DBC vs. AMB The method used to bond the conductive copper layers to the ceramic determines the interface's resistance to delamination. As summarized in Table 4.3, while DBC is the legacy standard for its cost-effectiveness, AMB has become the necessity for high-reliability SiC modules due to the superior chemical bonding provided by the active braze alloy.

4.4.1 Material Choices: Al_2O_3, AlN, and Si_3N_4

Three primary ceramic materials dominate the market, each representing a specific trade-off between thermal conductivity (k) and mechani-

Table 4.3: Comparison of DBC and AMB substrate metallization approaches for SiC power modules.

Attribute	DBC	AMB	Notes / reliability implication
Bonding method	Cu–O eutectic	Active braze	AMB uses Ti-based chemistry for non-oxide ceramics.
Ceramics used	Al_2O_3, AlN	Si_3N_4, AlN	AMB is ideal for low-reactivity Si_3N_4 surfaces.
PC robustness	Medium	High	AMB reduces delamination risk under large ΔT cycling.
Induced stress	Higher	Variable	Both benefit from symmetric, CTE-balanced stackups.
Cost	Low–Med.	Med.–High	Often justified by high-reliability automotive targets.

cal fracture toughness (K_{IC}):

- Al_2O_3 **(Alumina):** Inexpensive and widely available, but poorly suited for SiC. Its low thermal conductivity (20–30 W/m·K) and low fracture toughness lead to early "cratering" or cracking under the high ΔT_j of SiC modules.
- AlN **(Aluminum Nitride):** Selected for applications requiring maximum heat extraction due to its high conductivity (160–200 W/m·K). However, it is inherently brittle and susceptible to fatigue-induced cracking during the rapid thermal transients characteristic of EV traction.
- Si_3N_4 **(Silicon Nitride):** The current benchmark for reliability. While its conductivity (90–120 W/m·K) is lower than AlN, its exceptional fracture toughness and flexural strength allow for significantly thinner ceramic layers (e.g., 0.32 mm). This compensates for the lower k, resulting in a competitive overall thermal resistance (R_{th}) with vastly superior mechanical longevity.

The comparative properties of these materials are detailed in Table

4.4.

Table 4.4: Representative material properties of ceramic substrates for SiC modules.

Property	Al_2O_3	AlN	Si_3N_4
Thermal Cond. (W/m·K)	20–30	160–200	90–120
Fract. Toughness ($MPa{\cdot}m^{1/2}$)	3–4	3–4	5–7
Flexural Strength (MPa)	300–350	350–450	600–800
Cycling Endurance	Low	Medium	Very High
Typical Thickness (mm)	0.63	0.63	0.32

4.4.2 Copper Delamination and Metallization Fatigue

The dominant failure mode at the substrate level is the delamination of the copper tracks from the ceramic core. This is primarily driven by the Coefficient of Thermal Expansion (CTE) mismatch between the copper ($\approx 17\,\text{ppm/K}$) and the ceramic ($\approx 3\text{–}5\,\text{ppm/K}$).

As the module cycles between $T_{j,min}$ and $T_{j,max}$, the copper layers exert significant shear stress on the ceramic interface. These stresses are amplified in SiC modules by:

- **High dv/dt and Localized Heating:** Rapid switching creates intense, localized thermal gradients directly under the die.
- **Thick Metallization:** To improve thermal spreading and current carrying capacity, copper layers are often thickened (up to $600\,\mu\text{m}$). This increases the stored elastic energy at the interface, making the bond more prone to peeling.
- **Stiffness of Sintered Joints:** The use of high-stiffness sintered silver die-attach (as opposed to compliant solder) transfers more mechanical strain directly into the substrate metallization.

By pairing Si_3N_4 with the AMB process, the braze layer provides a compliant yet chemically robust interface that can withstand these stresses. Industry data suggests that Si_3N_4 AMB substrates can achieve 5–10× the cycling life of standard Al_2O_3 DBC, making them indispensable for the next generation of automotive traction inverters.

4.5 Interconnect Technologies

Interconnects are responsible for carrying high-density currents from the SiC MOSFET die to the module's external terminals. In SiC systems, the interconnect is no longer just a passive conductor; it is a critical component that dictates switching performance. Because SiC switches with extremely high di/dt and dv/dt rates, the parasitic inductance (L_σ) and resistance (R_σ) of the interconnect directly influence voltage overshoot ($V = L \cdot di/dt$), ringing, and electromagnetic interference (EMI).

As detailed in Table 4.5, the industry is currently transitioning from legacy wirebonding to advanced planar structures to keep pace with the performance potential of wide-bandgap materials.

Table 4.5: Comparison of interconnect technologies for SiC power modules.

Metric	Wirebonds (Al)	Clip bonds (Cu)	Impact on SiC
Parasitic Ind.	High ($\sim$10–15 nH)	Low (<2 nH)	Higher overshoot; increased EMI.
Thermal Life	Low	Very High	Key driver for long-term reliability.
Current Cap.	Med. (point)	Very High (planar)	Higher current and power density.
Mfg. Cost	Low	Medium	Trade-off for automotive scaling.

4.5.1 Aluminum Wirebonds: The Legacy Bottleneck

Wirebonding is a mature, ultrasonic-welding approach favored for its low cost and flexibility. However, the operational environment of SiC exposes several intrinsic failure modes:

- **Heel Cracking:** Repetitive thermal expansion causes the "heel" (the transition point where the wire meets the bond pad) to flex and fatigue.

- **Bond-Lift Off:** The Coefficient of Thermal Expansion (CTE) mismatch between the Aluminum wire (23 ppm/°K) and the SiC die (~4 ppm/°K) creates shear stress that eventually peels the bond from the pad interface.
- **Inductive Parasitics:** Multiple long wire loops create significant parasitic inductance, which leads to high-frequency ringing that can challenge the dielectric limits of the gate-oxide.

Standard Al wirebond lifetimes degrade sharply when junction temperatures consistently exceed 150–175°C, often failing to meet the rigorous 15-year service life required for automotive traction.

4.5.2 Copper Clip Bonds and Planar Interconnects

Copper clip bonding replaces thin wires with a solid, planar leadframe "bridge" soldered or sintered directly to the die surface. This transition from point-contacts to surface-contacts provides significant advantages:

- **Electromagnetic Optimization:** By reducing the loop area and utilizing a wide planar geometry, clip bonds can reduce parasitic inductance by over 80% compared to wirebonds.
- **Thermal Spreading:** The copper clip acts as a top-side heat spreader, allowing heat to dissipate from the top of the die in addition to the traditional path through the substrate.
- **Mechanical Resilience:** Copper clips provide a much higher cross-sectional area, drastically reducing current crowding and the resulting localized Joule heating.

4.5.3 Emerging Technologies and Hot-Spot Mitigation

The trend toward **Double-Sided Cooling (DSC)** has spurred the development of advanced interconnects such as flexible-film redistribution layers (RDLs) and sintered silver clips. A primary driver for these technologies is the elimination of interconnect-induced hot spots.

In wirebonded modules, current "crowds" at the bond-foot, creating localized temperature spikes 10–30 °C higher than the average junction temperature. As highlighted in Table 4.5, moving to copper-based planar interconnects ensures a uniform thermal profile across the die surface. This prevents the accelerated aging of the encapsulation material and die-attach layers, supporting the higher power densities required for modern inverters.

4.6 Heat-Spreading and Thermal Path Physics

The thermal management of SiC power modules is not merely a scaling of existing Silicon (Si) practices; it is a fundamental shift in heat-spreading dynamics. While SiC MOSFETs are more efficient, their significantly smaller die area for a given current rating creates a "thermal bottleneck" at the die-attach interface. In a typical SiC traction inverter, heat flux can reach 300 W/cm^2, a level that challenges the limits of conventional air- or liquid-cooled packaging.

4.6.1 Vertical vs. Lateral Heat Spreading

In traditional power electronics, the thermal path is often simplified as a 1D vertical flow. However, in SiC modules, **lateral heat spreading** within the metallization and ceramic layers is critical to reducing the effective thermal resistance (R_θ).

The heat follows a hierarchical extraction path:

$$\text{Junction } (T_j) \xrightarrow{\text{Vertical}} \text{Die attach} \xrightarrow{\text{Lateral spreading}} \text{Substrate} \xrightarrow{\text{Vertical}} \text{Baseplate} \xrightarrow{\text{Interface}} \text{TIM} \rightarrow \text{Cold plate} \tag{4.2}$$

Because the SiC die is small, the heat must "bloom" outward as it moves down the stack. This phenomenon, known as **constriction resistance**, implies that the first few hundred microns of the path (the die-attach and the top copper layer of the substrate) are the most critical. If these layers do not support rapid lateral spreading, the effective area of the subsequent layers (ceramic and baseplate) is underutilized, leading to localized hot spots that can exceed the junction's safe operating limit even if the average temperature remains low.

4.6.2 Thermal Conductivity of Key Packaging Materials

The efficiency of this spreading is determined by the thermal conductivity (k) of the constituent materials. Table 4.6 provides the physical constants required for accurate Finite Element Method (FEM) modeling of these paths.

Table 4.6: Thermal conductivities of materials in modern SiC power modules.

Material	k [W/m·K]	Thermal Functional Role
SiC Die	120–150	Primary heat source; lateral spreading
Sintered Ag	200–300	High-perf. die-attach; low R_{th}
Solder (SAC305)	50–70	Legacy die-attach; high R_{th} penalty
Al_2O_3 **Ceramic**	20–30	Insulator; poor thermal spreader
AlN Ceramic	160–200	High-performance spreader and insulator
Si_3N_4 **Ceramic**	90–120	High toughness; moderate conductivity
TIM (Grease)	3–8	Major bottleneck; subject to aging
Graphite Pad	10–20	High-stability automotive interface

4.6.3 Thermal Bottlenecks and Degradation

Experimental data consistently identifies the **Thermal Interface Material (TIM)** and the **Die Attach** as the two primary bottlenecks. While the ceramic substrate is a fixed resistance, the TIM and Die Attach are dynamic; they degrade over the mission profile. For instance, voiding in the die-attach increases localized R_θ, while "pump-out" of TIM grease increases the resistance at the baseplate-to-cold-plate interface. Managing these bottlenecks is the primary driver for moving toward advanced module topologies like Double-Sided Cooling.

4.7 Double-Sided Cooling and Advanced Module Topologies

As power densities push beyond the 200 W/cm^2 threshold, single-sided cooling reaches a point of diminishing returns. No matter how efficient the cold plate is, the single vertical path creates a steep temperature gradient that limits the peak current of the MOSFET. **Double-Sided Cooling (DSC)** addresses this by introducing a symmetric heat extraction path, effectively turning the power module into a "thermal sandwich."

4.7.1 Principles of DSC: Halving the Resistance

By placing a substrate and cooling structure on both the top (source/gate) and bottom (drain) of the SiC die, the total junction-to-case resistance ($R_{\theta jc}$) can be reduced by 30–50%. The heat extraction can be modeled as two parallel thermal resistors:

$$\frac{1}{R_{\theta,total}} = \frac{1}{R_{\theta,top}} + \frac{1}{R_{\theta,bottom}} \tag{4.3}$$

Beyond the absolute reduction in T_j, DSC offers a significant mechanical advantage: **Thermomechanical Symmetry:** In single-sided modules, the die tends to "bow" or warp due to the asymmetric CTE mismatch between the ceramic and the copper. In a DSC stack, the die is held between two similar structures, balancing the shear stresses and significantly extending the fatigue life of the interconnects.

4.7.2 The Necessity of Planar Interconnects

DSC is functionally impossible with traditional wirebonding. The height and fragility of wire loops prevent the application of a top-side cooling plate. Therefore, the transition to DSC is inherently a transition to **planar interconnects**, such as:

- **Copper Clips:** Solder- or sinter-bonded bridges that provide a flat top-surface.
- **Molybdenum (Mo) Spacers:** Often used to match the CTE of the die while providing the necessary vertical clearance for the gate structures.

In many automotive inverters, the interconnect and cooling system are co-designed; the copper clip that carries the source current simultaneously acts as the thermal conduit for the top-side heat sink.

4.7.3 Implementation Challenges: Pressure and Tolerances

While thermally superior, DSC modules are significantly more difficult to manufacture. The primary engineering hurdle is **Pressure Uniformity**. To ensure low thermal resistance, the dual cold plates must be clamped together with high force. However, any misalignment or thickness variation (tolerance stack-up) in the spacers, die, or substrates can create a localized pressure point that cracks the brittle SiC die.

Modern DSC designs mitigate this by using:

1. **Compliant TIMs:** High-performance pads that can accommodate micro-scale thickness variations.
2. **Integrated Manifolds:** Where the cooling fluid flows directly over the substrate metallization (Direct-Liquid Cooling), eliminating the baseplate and one TIM layer entirely.

4.8 Thermal Interface Materials (TIMs): Behavior and Degradation

Thermal interface materials bridge the microscopic gaps between nominally flat surfaces: typically between the module baseplate (or direct-cooled substrate) and the cooling plate. Even though TIM layers are thin—often 50–200 μm—their relatively low thermal conductivity and vulnerability to aging make them a dominant contributor to long-term thermal resistance drift in SiC power converters.

4.8.1 TIM Material Classes

Four material classes are most common in SiC modules:

1. **Silicone-based greases** with ceramic or metallic fillers: easy to apply and rework, but susceptible to pump-out and dry-out at high temperature.
2. **Phase-change materials (PCMs):** solid at room temperature, soften or melt near operating temperature; good wetting and stable thickness, but limited long-term performance at very high temperature.
3. **Graphite sheets and pads:** increasingly popular in automotive traction inverters due to stable performance, relatively high in-plane conductivity, and resistance to pump-out.
4. **Metallic TIMs** (solder or sintered metal foils): provide very low thermal resistance and excellent stability but require more complex assembly and are often non-reworkable.

The choice balances thermal performance, manufacturability, reworkability, mechanical reliability, and cost.

4.8.2 TIM Degradation Mechanisms

TIM layers degrade over time through several mechanisms:

- **Pump-out:** Repeated expansion and contraction of the module relative to the cooling plate causes the TIM to migrate laterally, thinning the layer directly beneath the die and thickening it near edges.
- **Void formation:** Mechanical cycling and material fatigue produce air gaps, especially at die corners and substrate edges, significantly increasing local thermal resistance.
- **Dry-out and oil bleed (for greases):** Volatile components evaporate or bleed out, leaving a more rigid, less conductive network of filler particles.
- **Particle segregation:** Filler particles can separate from the carrier fluid, leading to inhomogeneous conductivity and localized hot spots.

Experimental data from long-duration cycling on EV-grade modules show that a 10–20 % increase in module-level $R_{\theta jc}$ over life is not uncommon when TIM degradation is not aggressively managed. In one sense, TIM aging acts as an accelerator for all other failure mechanisms: as $R_{\theta jc}$ slowly increases, junction temperatures rise, die-attach fatigue accelerates, substrate stresses sharpen, and interconnect fatigue worsens.

4.8.3 TIM Design and Assembly Considerations

To mitigate TIM-related reliability problems, module and system designers commonly:

- minimize TIM thickness consistent with flatness tolerances and electrical isolation,
- ensure surface planarity and roughness within controlled limits to reduce local voids,

- maintain uniform clamping pressure across the module footprint,
- select TIMs with proven cycling stability at the target temperature range.

In double-sided cooled modules, these requirements apply simultaneously to two interfaces, amplifying the importance of careful mechanical co-design.

4.9 Transient Thermal Impedance and Electrothermal Modeling

Static thermal resistance ($R_{\theta jc}$) characterizes steady-state behavior, but SiC converters frequently operate under conditions where the thermal response is dynamic: short-circuit events, overloads, rapid torque transitions, and high-frequency pulsed operation. In these regimes, the *transient thermal impedance* $Z_\theta(t)$ is the relevant quantity.

4.9.1 Foster and Cauer Thermal Networks

Transient thermal behavior is often approximated by RC networks:

- **Foster networks** are purely mathematical curve fits of $Z_\theta(t)$. They match measured thermal impedance but lack direct correspondence to physical layers.
- **Cauer networks** map each RC element to an actual physical layer (die, die attach, substrate, baseplate, TIM, etc.), providing a more intuitive interpretation and easier modification when the stack changes.

For packaging design and reliability analytics, the Cauer form is generally preferred.

4.10 Short-Circuit and Overload Conditions

During a short-circuit (SC) event, a SiC MOSFET must dissipate a massive amount of energy—often 10 to 50 times its nominal rating—within a window of only a few microseconds. Because the time scale is so short, the heat does not have time to escape the die and enter the substrate; instead, it stays localized within the SiC drift layer and source metallization. Accurate transient thermal modeling is required to:

- Estimate the instantaneous junction temperature rise (ΔT_j) during a fault.
- Determine the critical short-circuit withstand time ($t_{\mathrm{SC,max}}$).
- Calibrate "desaturation" (Desat) protection or current-sense thresholds in the gate driver.
- Evaluate multi-pulse robustness during repetitive overload cycles.

4.10.1 Transient Thermal Calculation: Duhamel's Integral

If the power dissipation $P_{\mathrm{SC}}(t)$ varies during the short-circuit event (e.g., due to the temperature-dependent reduction in transconductance or V_{DS} ringing), the junction temperature rise $\Delta T_j(t)$ is calculated using the convolution of the power profile and the thermal impulse response:

$$\Delta T_j(t) = \int_0^t P_{\mathrm{SC}}(\tau) \cdot \frac{dZ_\theta(t-\tau)}{d(t-\tau)}\, d\tau \tag{4.4}$$

where $Z_\theta(t)$ is the time-domain thermal impedance curve, typically represented by a multi-stage Foster or Cauer RC network. For constant power pulses (P_{SC}), Equation 4.4 simplifies to the familiar $\Delta T_j(t) = P_{SC} \cdot Z_\theta(t)$.

4.10.2 The Adiabatic Limit and the $\sqrt{t}$ Relationship

For ultra-short pulses ($t < 10\,\mu$s), the system operates in the **adiabatic regime**. In this state, heat conduction into the die-attach is negligible, and the thermal impedance is dominated by the semi-infinite diffusion of heat within the SiC material itself. Under these conditions, the thermal impedance follows a "square-root-of-time" behavior:

$$Z_\theta(t) \approx \frac{2}{\sqrt{\pi \cdot k \cdot \rho \cdot C_p} \cdot A} \cdot \sqrt{t} \tag{4.5}$$

where k is thermal conductivity, ρ is density, C_p is specific heat, and A is the active die area. This relationship is critical for SiC reliability because the smaller die size of SiC MOSFETs (relative to Si IGBTs) results in a much faster temperature rise for the same SC energy, often leading to gate-oxide rupture or aluminum metallization melting if the fault is not cleared within 2–5 μs.

4.10.3 Failure Mechanisms under Short-Circuit

Experimental results indicate that SiC MOSFETs typically fail due to one of two mechanisms:

1. **Thermal Runaway:** High temperatures trigger intrinsic carrier generation, leading to uncontrollable leakage current and eventual melting of the crystalline structure.

2. **Gate-Oxide Degradation:** Even if the device survives the pulse, the extreme T_j transients can cause thermomechanical stress that leads to latent damage in the SiO_2 interface, manifesting as a shift in threshold voltage (V_{th}) or increased gate leakage (I_{GSS}).

4.10.4 Coupled Electrothermal Simulation

For high-power SiC modules, purely electrical simulation is insufficient to verify safe operation. Electrothermal co-simulation is increasingly used to:

- capture the dependence of $R_{DS(on)}$ and threshold voltage on temperature,

- estimate current redistribution between parallel chips as they heat unevenly,

- predict hot-spot formation at the die and module level,

- link mission-profile power losses to long-term thermal cycling damage.

These simulations form the basis for physics-of-failure (PoF) modeling.

4.11 Packaging Aging Across Layers

Even when the SiC die itself remains electrically robust, the package surrounding it gradually ages under combined thermal, mechanical, and electrical stress. A realistic reliability model must therefore treat the module as a stack of interacting layers, each subject to different failure mechanisms.

4.11.1 Layer-by-Layer Aging Mechanisms

From bottom to top, typical degradation processes include:

- **Cooling plate and baseplate:** corrosion, erosion from coolant flow, mechanical fatigue under vibration and pressure cycling.
- **TIM layer:** pump-out, voiding, dry-out, and particle segregation, leading to increased $R_{\theta jc}$ and higher T_j.
- **Ceramic substrate (DBC/AMB):** cracking due to thermal bending, copper delamination, and fatigue at the metal–ceramic interface.
- **Die-attach (solder or sintered Ag):** crack initiation at die corners, void growth, and loss of adhesion under cyclic shear strain.
- **Interconnects (wirebonds or clips):** heel cracking, lift-off, or local delamination under thermal and electrical cycling.
- **Encapsulation and molding compounds:** delamination, moisture ingress, and partial-discharge-related erosion, especially at high dv/dt.

4.11.2 Feedback Between Layers

A key insight from multi-physics reliability studies is that packaging failures rarely occur in isolation. Instead, they form a feedback loop:

1. TIM degradation increases thermal resistance $\rightarrow$ higher T_j and larger ΔT_j per cycle.
2. Larger ΔT_j accelerates die-attach fatigue and substrate cracking $\rightarrow$ local hot spots develop as cracks reduce effective conduction area.

3. Hot spots further elevate local temperatures and exacerbate interconnect fatigue → wirebond or clip failures increase local current crowding.

4. Elevated local losses lead to even higher temperatures, feeding back into step (1).

This feedback loop explains why field failures in SiC modules often appear sudden after a long period of seemingly benign operation: *multiple degradation processes quietly accumulate until a tipping point is reached.*

4.12 Design Guidelines and Best Practices

Synthesizing the experimental and modeling results across the literature, several robust guidelines consistently emerge for SiC power-module packaging:

- **Die attach:** Use sintered-Ag die attach for high-cycling, high-ΔT_j applications. Limit solder die attach to cost-sensitive, moderate-stress converters or auxiliary systems.
- **Substrates:** Prefer Si_3N_4-based DBC or AMB substrates for traction, aerospace, and other high-reliability use cases; AlN is acceptable when thermal performance is critical and cycling is moderate.
- **Interconnects:** Avoid wirebonds in high-frequency, high-power SiC modules. Adopt copper clips, ribbons, or planar interposers to reduce parasitics and improve cycling lifetime.
- **Cooling topology:** Employ double-sided cooling when power densities exceed 100–150 W/cm^2, or when mission profiles impose large ΔT_j cycles. Co-design the cooling plate, module housing, and TIM interfaces.
- **TIM management:** Carefully control TIM thickness, clamping pressure, and surface flatness. Use TIM materials with demonstrated stability under the relevant temperature and cycling range.
- **Thermal modeling:** Develop Cauer-based thermal networks early in the design cycle and validate them experimentally. Use

electrothermal co-simulation to verify short-circuit robustness and transient overload behavior.

- **Mission-profile analysis:** Always map the target application mission profile into temperature cycles using Rainflow counting, then apply lifetime models and PoF analysis to verify margin relative to required lifetime (e.g., 15 years for PV inverters, 15–20 years or 300k km for EV traction).

These practices have been shown to extend module life by factors of 3–10 compared with legacy packaging, especially when all layers of the packaging stack are co-optimized rather than treated independently.

4.13 Conclusion

This chapter has developed a comprehensive view of SiC power-module packaging and thermal management, emphasizing that *packaging—not the semiconductor die—often sets the practical limit on SiC performance and lifetime.* The main conclusions are:

- SiC devices impose higher switching speeds, higher junction temperatures, and higher power densities than Si IGBTs, forcing a shift to new packaging paradigms.

- Die-attach technologies, substrate ceramics, and interconnect strategies strongly influence both thermal and mechanical performance. Sintered Ag, Si_3N_4 substrates, and clip bonds have emerged as key enablers for high-reliability SiC modules.

- Advanced cooling architectures, especially double-sided cooling, are necessary to fully exploit SiC's thermal capability and to reduce thermal gradients and cycling stress.

- TIM layers, although thin, are a major source of long-term thermal-resistance increase. Their degradation amplifies all other failure mechanisms by raising junction temperature.

- Packaging aging is a multi-layer, multi-physics problem: die attach, substrate, interconnects, and TIM interact through coupled thermal, mechanical, and electrical feedback loops.

- Mission-profile–based lifetime models, combining Rainflow cycle counting, Coffin–Manson-type fatigue laws, Arrhenius acceleration, and PoF co-simulation, are essential to bridge the gap between laboratory tests and field reliability.

With this foundation, subsequent chapters focus on the **gate-drive and system-integration aspects** of SiC technology. Chapter 5 examines gate-driver architectures, dv/dt control, protection strategies, and EMI considerations, and shows how they must be co-designed with packaging and thermal-management choices to achieve robust, high-efficiency SiC power systems.

Chapter 5

Gate-Drive Design, Protection, and Control for SiC MOSFETs

5.1 Introduction

This chapter consolidates the *Control & Gate-Drive Master* definitions for SiC switching: Miller plateau behavior, gate-loop inductance, switching-shaping strategies, and protection timing. Device-level origins of capacitance and charge storage are introduced in Chapter 2, and reliability consequences of overstress (including short-circuit and diode degradation) are consolidated in Chapter 3. Chapter 6 applies the design rules of this chapter to complete converters, including EMI filter co-design and system-level integration.

5.2 Gate-Voltage Requirements and Device Behavior

SiC MOSFETs exhibit higher threshold voltages and more sensitive gate-oxide electric-field distributions than silicon MOSFETs. Typical recommended drive voltages are:

- **Turn-on:** +15 to +20 V (depending on manufacturer and trench vs. planar structure),

- **Turn-off:** –3 to –5 V for fast-switching modules (to suppress false turn-on),
- **UVLO levels:** 10–12 V for turn-on undervoltage protection.

Driving the gate well above the threshold is critical because SiC threshold voltage drifts under bias temperature instability (BTI) and charge trapping (Chapter 3). If V_{GS} is marginal at turn-on, conduction losses may increase substantially over time as the effective threshold shifts. Conversely, excessive V_{GS} accelerates gate-oxide wear-out and reduces TDDB margin. Therefore, maintaining a tightly regulated gate rail—typically using Kelvin source sensing and low-impedance isolated supplies—is essential.

A practical design compromise for many automotive-grade devices is:

- $V_{\text{GS,on}} = +15$ V (or $+18$ V for devices specifically rated),
- $V_{\text{GS,off}} = -3$ to -4 V,
- UVLO thresholds set such that the device is guaranteed off below $\approx$ 10–12 V.

These values balance conduction efficiency, oxide lifetime, and immunity to false turn-on.

5.3 Gate-Charge Dynamics, Miller Plateau, and Gate-Loop Inductance

This chapter is the single reference for practical gate-drive definitions and layout-centric switching behavior. The material-physics origin of capacitances is introduced in Chapter 2, while the reliability consequences of overstress are treated in Chapter 3.

5.3.1 Gate-Charge Dynamics and the Miller Plateau

The gate-charge profile of a SiC MOSFET is a primary determinant of switching energy (E_{sw}) and attainable switching frequency (f_{sw}). While SiC devices generally feature a lower total gate charge (Q_g) than Silicon IGBTs of similar ratings, the gate-charge evolution follows the three distinct physical phases described below and illustrated in Fig. 5.1:

1. **Phase I: Charging C_{gs} and $C_{gd}(V_{DS,high})$:** The gate voltage V_{GS} rises from zero (or a negative off-state voltage) toward the threshold voltage V_{th}. In this region, the device remains in the cut-off state.

2. **Phase II: The Miller Plateau:** Once V_{GS} reaches the plateau voltage (V_{GP}), the drain current I_D is fully established. The gate current I_g is now consumed entirely by the discharge of the highly non-linear gate-drain capacitance (C_{gd}), while V_{DS} falls. Consequently, V_{GS} remains nearly constant.

3. **Phase III: Final Gate Charging:** After V_{DS} has fallen to its on-state value ($V_{DS,on}$), V_{GS} rises from the plateau toward the driven gate voltage ($V_{GS,on}$), further charging C_{gs} and C_{gd} to minimize the on-resistance ($R_{DS,on}$).

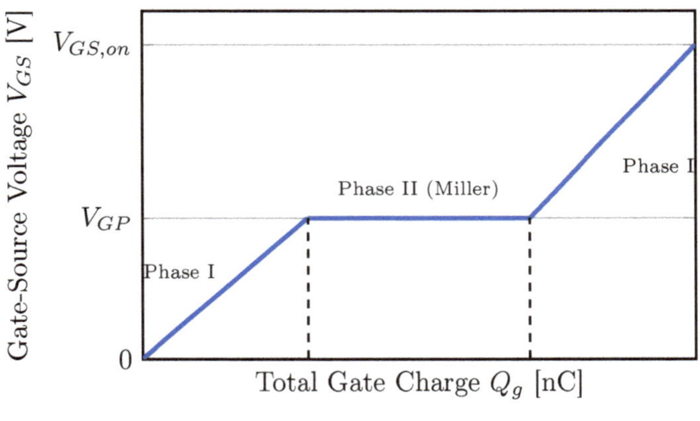

Figure 5.1: Conceptual gate-charge curve of a SiC MOSFET. The duration of Phase II is dominated by the transfer of charge Q_{gd}.

During the Miller plateau (Phase II), the gate current i_g is defined by the rate of change of the gate-drain charge:

$$i_g = \frac{dQ_{gd}}{dt} = C_{gd}(V_{DS}) \cdot \frac{dV_{GD}}{dt} \tag{5.1}$$

Since V_{GS} is constant during the plateau, $dV_{GD}/dt \approx -dV_{DS}/dt$. Therefore, the slew rate of the drain-source voltage is governed by the gate current and the non-linear reverse transfer capacitance:

$$\frac{dV_{DS}}{dt} \approx -\frac{i_g}{C_{gd}(V_{DS})} \tag{5.2}$$

This relationship underscores why the dV_{DS}/dt of SiC MOSFETs—often exceeding 50–100 V/ns—is highly sensitive to the external gate resistance ($R_{g,ext}$) and the driver's peak current capability. To mitigate the resulting EMI and voltage overshoot without sacrificing efficiency, high-performance SiC packages utilize Kelvin-source pins to decouple the power-loop di/dt from the gate-control loop.

5.3.2 Power-Loop and Gate-Loop Parasitics

Switching waveforms are significantly shaped by parasitic inductances in the power loop and gate loop. Even small parasitic inductances (1–3 nH) can generate tens of volts of overshoot during high di/dt transitions:

$$V_L = L_{\text{parasitic}} \cdot \frac{di}{dt}.$$

With SiC di/dt values often reaching 5 A/ns, overshoot can exceed 25 V for a 5 nH loop inductance. High-frequency conducted EMI in SiC-based converters is likewise dominated by common-mode currents excited by rapid switching transitions and the parasitic capacitances of the module and heat sink.

The most critical parasitic is the common-source inductance (CSI). CSI couples the drain current to the gate-source voltage, modifying the effective gate-drive waveform:

$$V_{GS,eff} = V_{drive} - L_{CS} \cdot \frac{di}{dt}.$$

High CSI reduces switching speed, increases switching losses, and may lead to false turn-on in half-bridge configurations if the coupled voltage is large enough to cross the threshold. Module and PCB layouts for SiC must therefore be co-designed with the gate driver, as will be elaborated in later Chapters.

5.4 Switching Dynamics: Charge, Slew Rates, and Parasitics

Much of the *underlying* switching behavior of SiC MOSFETs—including charge partitioning, the Miller plateau, output-capacitance nonlinearity, and the origin of high dv/dt and di/dt—is developed in Chapter 2. In practice, the same physics becomes a *design problem* once package and layout parasitics translate device current and voltage slew rates into overshoot, ringing, common-mode currents, and false turn-on.

For the remainder of this chapter, we therefore treat switching dynamics from the *gate-driver viewpoint*: what the driver can control, what it cannot, and how to translate device data-sheet charge parameters into robust, repeatable switching behavior. Packaging-induced parasitics and their coupling paths are treated in Chapter 4, and will be referenced as needed rather than re-derived here.

- **What the driver directly shapes:** gate current waveform, turn-on/turn-off resistance (static or dynamic), and clamp/negative-bias strategy.
- **What the power stage contributes:** power-loop inductance, common-source inductance, and high-frequency capacitance coupling (module and PCB layout).
- **What to validate experimentally:** peak V_{DS}, ringing frequency/damping, gate-voltage margin during Miller, and EMI or common-mode current signatures.

The next sections build a practical toolkit for dv/dt and di/dt management, Miller/false turn-on suppression, and short-circuit protection constraints specific to SiC devices.

5.5 dv/dt and di/dt Control

The driver must carefully regulate switching slew rates to avoid excessive overshoot and EMI, while still preserving the efficiency benefits of SiC. The primary control mechanism is the gate resistance R_G, which can be separated into:

- **Turn-on gate resistance** $R_{G,\mathrm{on}}$,
- **Turn-off gate resistance** $R_{G,\mathrm{off}}$,

often implemented using diode-resistor networks. Larger $R_{G,\mathrm{on}}$ reduces *di/dt* at turn-on and can substantially reduce voltage overshoot and reverse-recovery stress in the complementary device. Larger $R_{G,\mathrm{off}}$ suppresses ringing and prevents voltage oscillation during turn-off but increases turn-off losses.

Advanced gate drivers support programmable or adaptive gate shaping, including:

- separate $R_{G,\text{on}}$ and $R_{G,\text{off}}$,
- multi-level gate control (intermediate voltage steps),
- current-source gate driving,
- digitally programmed slew-rate control.

These techniques allow designers to tailor dv/dt and di/dt to EMI, insulation, and reliability constraints without unnecessarily sacrificing efficiency.

5.6 Miller Effect and False Turn-On

The Miller capacitance C_{GD} couples drain-voltage transitions into the gate. During high dv/dt, the induced Miller current is:

$$i_{\text{Miller}} = C_{\text{GD}} \frac{dV_{\text{DS}}}{dt}.$$

If this current flows through the gate resistance and raises V_{GS} above threshold, the complementary device may partially turn on during the opposite switch transition. This risks shoot-through, large reverse-recovery currents, and severe overshoot.

False turn-on is particularly problematic in:

- half-bridge and full-bridge topologies with hard switching,
- paralleled MOSFETs with unequal parasitics,
- high-temperature operation where V_{th} may shift downward.

Mitigation strategies include:

- negative turn-off gate bias (–3 to –5 V),
- split gate-source resistor networks (dedicated R_G for each device),
- Kelvin source connections to reduce common-source inductance,
- active Miller clamps (low-impedance path to ground during dv/dt),
- careful reduction of C_{GD} and C_{oss} in device design (seen in modern trench MOSFETs).

Device datasheets often specify maximum recommended dv/dt values that assume proper use of these techniques.

5.7 Short-Circuit Behavior and Gate-Driver Constraints

Short-circuit event physics (overshoot, droop, runaway) and reliability implications are consolidated in Chapter 3. Here the emphasis is on protection implementation: sensing, timing, and gate-control strategies that keep the device within its validated SOA.

5.8 Summary of Gate-Drive Fundamentals

The foundational requirements for SiC MOSFET gate-drive design can be summarized as follows:

- Gate voltage levels must be tightly regulated to balance conduction performance and oxide reliability.
- Switching speed is limited primarily by package parasitics and EMI constraints, not by semiconductor physics.
- dv/dt and di/dt must be controlled to avoid overshoot, EMI, and device overstress.
- The Miller effect is a central concern, requiring careful turn-off design and Kelvin source routing to prevent false turn-on.
- Short-circuit protection must respond much faster than in silicon-based systems due to the rapid thermal buildup in SiC devices.

The following sections build on these fundamentals to develop concrete protection schemes, advanced gate-control methods, and practical layout guidelines.

5.9 Protection Requirements for SiC MOSFETs

SiC MOSFETs offer exceptional switching performance, but their robustness window is narrower than that of silicon IGBTs. Their short-circuit withstand times t_{SC} are typically in the range of 2–5 µs for automotive-grade 650–1200 V devices and may be even shorter for high-power-density trench MOSFETs. Because SiC thermal time constants are extremely small—dominated by only a few micrometers of active area—the device can transition into thermal runaway within microseconds.

This places stringent requirements on the gate driver and protection circuitry:

- detect faults within 200–500 ns,
- initiate soft turn-off to control voltage overshoot,
- isolate high dv/dt events from the control domain,
- avoid false trips during normal switching transients.

Three major protection categories are used in SiC gate drivers:

1. **Short-circuit protection** (DESAT or current-based detection),
2. **Overvoltage protection** (active clamping and Miller suppression),
3. **Overcurrent and shoot-through protection** (cycle-by-cycle limiting and dead-time management).

The following subsections present each mechanism in detail, emphasizing practical implementation and constraints imposed by package inductance and switching speed.

5.10 Short-Circuit Behavior and Safe Operating Area (SOA)

Short-circuit event physics (overshoot, droop, runaway) and reliability implications are consolidated in Chapter 3. Here the emphasis is on protection implementation: sensing, timing, and gate-control strategies that keep the device within its validated SOA.

5.11 Desaturation (DESAT) Detection and Driver Response

Desaturation (DESAT) detection is the primary active protection mechanism employed by modern gate drivers to shield SiC MOSFETs from short-circuit (SC) failure. From the perspective of the gate driver, the "desaturation" of the MOSFET refers to the point where the device leaves the linear (ohmic) region and enters the saturation (active) region. In this state, the V_{DS} rises rapidly while the current is limited only by the device's transconductance, leading to extreme power dissipation.

5.11.1 The DESAT Monitoring Network

A robust gate driver implementation requires a peripheral sensing network to translate the high-voltage drain signal into a logic-level comparator input. The typical driver-side components include:

- **High-Voltage Blocking Diode (D_{desat}):** Provides isolation between the high-voltage drain (up to 1.2 kV or more) and the low-voltage driver pins during the off-state.
- **Internal Blanking Current (I_{blank}):** A precision current source (typically 250–500 µA) within the driver that charges an external blanking capacitor (C_{blank}).
- **DESAT Comparator:** An integrated comparator with a fixed reference voltage ($V_{th,DESAT}$), usually set between 6 V and 9 V.

During conduction, if $V_{DS} + V_{F,diode}$ exceeds the comparator threshold, the driver identifies a fault. Because SiC MOSFETs have a significantly lower $R_{DS,on}$ and a "sharper" transition into saturation than Silicon IGBTs, the driver's ability to distinguish between a heavy load and a true short circuit is a critical design constraint.

5.11.2 Blanking Time and Detection Latency

The gate driver must ignore the high V_{DS} present during the initial turn-on transient to avoid "false tripping." This interval is defined as the **blanking time** (t_{blank}).

In Silicon-based systems, t_{blank} could safely reside in the 2–5 µs range. However, for SiC MOSFETs, the reduced short-circuit withstand time (SCWT) necessitates an aggressive reduction in detection latency. The effective blanking time is typically governed by:

$$t_{blank} = \frac{C_{blank} \cdot V_{th,DESAT}}{I_{blank}} \tag{5.3}$$

For SiC applications, the driver is often configured for $t_{blank} <$ 500 ns. If the blanking time is too long, the die may reach critical temperatures before the driver can intervene; if it is too short, the driver may trip during the Miller plateau when V_{DS} has not yet fully collapsed.

5.11.3 Kelvin Source Referencing for Noise Immunity

Accurate DESAT sensing in high-speed SiC circuits is highly sensitive to parasitic ground bounces. At di/dt rates exceeding 5 A/ns, even a few nanohenries of common-source inductance (L_s) can create a voltage offset:

$$V_{error} = L_s \cdot \frac{di}{dt} \tag{5.4}$$

If the DESAT return path is referenced to the power source (high-current path), this V_{error} will be added to the perceived V_{DS}, leading to premature fault triggering. Consequently, high-performance SiC drivers **must** be referenced to a dedicated Kelvin source pin. This ensures the driver's local ground is "riding" the same potential as the die's source metallization, providing a clean differential measurement of the channel voltage.

5.11.4 Fault Handling: Soft Turn-Off (STO)

Once the DESAT threshold is breached, the gate driver must not execute a standard "hard" turn-off. Abruptly interrupting a short-circuit current (which may be several kA) would induce a lethal voltage overshoot ($V_{OS} = L_{loop} \cdot di/dt$) across the MOSFET.

Instead, the driver initiates a **Soft Turn-Off (STO)** sequence:

1. **Current Limiting:** The driver partially reduces the gate-source voltage (V_{GS}) to a value just above the threshold, which significantly reduces the peak short-circuit current while keeping the device in a controlled state.

2. **Controlled Discharge:** The gate is slowly discharged through a high-impedance internal resistor (R_{STO}), extending the fall time of the current (t_f) over several microseconds.

3. **Negative Bias and Latching:** Once the current has safely decayed, the gate is pulled to a negative off-state (e.g., –5 V),

and the fault is latched, requiring a system-level reset before the inverter can restart.

5.12 Current-Based Overcurrent Protection (OCP)

While DESAT is the standard for catastrophic short-circuit protection, many high-reliability systems incorporate secondary Current-Based Overcurrent Protection (OCP). Unlike DESAT, which monitors the state of the MOSFET channel, OCP directly measures the load or phase current using sensors such as:

- **Low-Inductance Shunts:** Preferred for precision but can introduce unwanted parasitic inductance into the power loop.
- **Hall-Effect Sensors:** Provide galvanic isolation and DC-to-AC sensing but may suffer from limited bandwidth in high-frequency SiC applications.
- **Rogowski Coils:** Offer excellent di/dt sensing and high bandwidth, though they cannot detect static DC faults.

The primary advantage of OCP is its **immunity to V_{DS} ringing**. During the hard-switching transients of SiC MOSFETs, the drain-source voltage oscillates significantly; OCP remains unaffected by these voltage swings, providing a cleaner signal for overload detection. However, OCP is typically slower than DESAT—with response latencies ranging from 300 ns to over 1 µs—due to the propagation delays in the analog filtering and signal conditioning chain. In a modern "defense-in-depth" architecture, DESAT is tuned to catch "hard" short circuits (Fault Type I), while OCP manages overloads and repetitive current-limiting events (Fault Type II).

5.13 Soft Turn-Off (STO) Design

Soft Turn-Off (STO) is a mandatory fail-safe mechanism. When a fault is detected, the current flowing through the MOSFET can be several times higher than the nominal rating. Abruptly interrupting this current would trigger a massive voltage overshoot (V_{OS}) dictated by the loop inductance:

$$V_{OS} = V_{bus} + L_{loop} \cdot \left| \frac{di}{dt} \right| \tag{5.5}$$

To prevent dielectric breakdown of the SiC oxide or the encapsulation, the gate driver must execute a controlled, high-impedance discharge.

Design Trade-offs The choice of the soft turn-off resistance ($R_{G,STO}$) is a balance between peak voltage and thermal energy. If $R_{G,STO}$ is too small, the V_{OS} remains dangerously high. If it is too large, the MOSFET stays in the linear region too long, leading to excessive Joule heating and potential thermal runaway during the turn-off event itself.

Sophisticated drivers employ **Two-Level Turn-Off (2LTO)**, which initially drops V_{GS} to an intermediate "plateau" (e.g., 8 V) to quickly cap the peak current, followed by a slower ramp to the off-state. This "gentle" extraction significantly reduces the stress on the package insulation and the device's guard rings.

5.14 Active Clamping and Overvoltage Protection

Active clamping acts as a dynamic safety valve, modifying the gate voltage if V_{DS} exceeds a critical threshold (typically 80–90% of the BV_{DSS}). In SiC design, gate-side active clamping is favored over passive drain-side snubbers because it utilizes the MOSFET's own channel to dissipate the transient energy.

When the drain voltage spikes, a TVS (Transient Voltage Suppressor) string conducts current back into the gate, slightly turning the MOSFET back on. This increases the channel's conductivity just enough to slow down the dv/dt, effectively "clamping" the voltage at the TVS breakdown level.

- **Stability:** The clamp must be coordinated with the DESAT blanking time to ensure the protection loops do not oscillate against one another.

- **Reliability:** While SiC is inherently rugged, repeated active clamping events can lead to threshold voltage (V_{th}) shifts over time due to high-energy carrier injection into the gate oxide.

5.15 Negative Gate Bias and False Turn-On Prevention

The low threshold voltage ($V_{th} \approx 2$–3 V) and high dv/dt characteristic of SiC MOSFETs make them highly susceptible to **parasitic turn-on**. During a switching event in a half-bridge, the "Miller current" ($I_{Miller} = C_{gd} \cdot dv/dt$) flows through the gate resistor of the off-state device. If the resulting voltage drop exceeds V_{th}, the device turns on momentarily, leading to shoot-through.

To provide a sufficient safety margin, designers utilize a negative off-state bias, typically between -3 V and -5 V. This shifts the starting point of the gate voltage further away from the "danger zone" of the threshold.

However, negative bias is not "free." It requires bipolar isolated power supplies and introduces the risk of "Bias Temperature Instability (BTI)". Excessive negative stress can cause V_{th} to drift over the device's lifetime, potentially increasing $R_{DS,on}$ or affecting current sharing in paralleled modules.

5.16 Shoot-Through Protection and Dead-Time Engineering

Dead-time engineering is the art of ensuring that the high-side and low-side switches are never "on" at the same time. In SiC systems, the dead-time calculation must account for the high speed of the devices while leaving room for propagation delay skews:

$$t_{dead} \geq (t_{off,max} - t_{on,min}) + t_{skew} + t_{margin} \tag{5.6}$$

Unlike Silicon IGBTs, where dead times are often > 1 µs, SiC allows for dead times as low as 50–300 ns. However, excessively long dead times are detrimental in SiC because the "body diode" (or the third-quadrant channel conduction) has a higher forward voltage drop ($V_F \approx 3$–5 V) than Silicon. Long dead times thus lead to increased conduction losses and higher heat generation. Adaptive dead-time control in the controller can optimize this window in real-time, balancing safety against efficiency.

5.17 Protection Timing Coordination

The effectiveness of a SiC protection suite depends entirely on the chronological coordination of events. A failure in timing is as catastrophic as a failure in sensing.

1. $t = 0$ **(Fault Occurs):** Current begins to rise at several kA/µs.
2. **Blanking Window:** The driver ignores the V_{DS} for 200–400 ns to allow the Miller plateau to pass.
3. **Detection:** DESAT or OCP triggers. The driver must filter out noise spikes (usually a 50–100 ns deglitch filter).
4. **Response Selection:** If the fault is severe, the driver bypasses the standard PWM and engages the STO path.
5. **Quenching:** The STO ramp takes 1–5 µs to bring the current to zero safely.
6. **Isolation:** The fault is latched, and the system-level controller is notified to halt the entire inverter.

5.18 Summary of Protection Concepts

The protection mechanisms required to operate SiC MOSFETs safely in high-performance power converters can be summarized as follows:

- Short-circuit withstand times are extremely short, requiring fault detection within hundreds of nanoseconds.
- DESAT remains the fastest and most effective protection method, but requires careful blanking, Kelvin reference routing, and coordination with soft turn-off.
- Soft turn-off is essential to avoid voltage overshoot during fault clearing and to protect module insulation.
- Active clamping and negative gate bias mitigate the effects of parasitic inductance and Miller coupling.
- Shoot-through protection and precise dead-time engineering are mandatory in high-switching-frequency SiC half-bridges.

- Protection strategies must be co-designed with layout, gate-loop inductance, and device-specific SOA characteristics.

The next sections build on these concepts to introduce advanced gate-control techniques and practical implementation details.

5.19 Advanced Gate Control and Switching-Shaping Techniques

As Silicon Carbide (SiC) MOSFETs push switching speeds into the tens or even hundreds of volts per nanosecond (V/ns), the limitations of "static" gate driving become apparent. While legacy Silicon IGBTs were often driven with a simple fixed resistor, SiC devices operating at high frequencies ($>$ 100 kHz) and high voltages ($>$ 800 V) require intelligent gate-shaping to manage the electromagnetic and thermomechanical side effects of extreme di/dt and dv/dt.

Advanced gate control aims to decouple the conflicting requirements of efficiency and reliability. By dynamically shaping the gate current or voltage, designers can minimize switching losses while simultaneously suppressing EMI, preventing parasitic turn-on, and ensuring balanced current sharing in paralleled modules.

5.19.1 Gate-Loop Dynamics and the Need for Active Control

The gate loop is the high-frequency control path that dictates the transient behavior of the MOSFET. In SiC systems, this loop is highly sensitive to parasitics. The dynamics are governed by a complex interaction between:

- **Driver Impedance:** The internal "push-pull" resistance of the gate driver IC.

- **Total Gate Resistance:** The sum of the internal device resistance ($R_{G,\text{int}}$) and the external board-level resistance ($R_{G,\text{ext}}$).

- **Common-Source Inductance (L_{CS}):** The most critical parasitic, which creates a negative feedback voltage ($V_L = L_{\text{CS}} \cdot di/dt$) that opposes the gate-drive signal during turn-on, effectively slowing the device.

- **Non-linear Capacitances:** Specifically the voltage-dependent Miller capacitance (C_{gd}), which varies by orders of magnitude during a single switching cycle.

Active gate control allows the system to switch "fast when it can, and slow when it must," bypassing the rigid trade-off between switching losses (E_{on} and E_{off}) and voltage ringing.

5.19.2 Split Gate Resistance for Asymmetric Switching Control

The first step toward advanced control is the implementation of asymmetric gate resistors ($R_{G,\text{on}} \neq R_{G,\text{off}}$). This is typically achieved using a diode-resistor network that provides independent paths for the charging and discharging currents.

- **Turn-On ($R_{G,\text{on}}$):** Usually set higher than the turn-off resistance to constrain the di/dt and limit the reverse-recovery current spike of the opposing free-wheeling diode.
- **Turn-Off ($R_{G,\text{off}}$):** Usually set lower to accelerate the removal of gate charge, minimizing the "current tail" and preventing Miller-induced false turn-on.

Beyond a simple split, designers may use **stepped-resistor networks**. For example, a small capacitor in parallel with a portion of $R_{G,\text{on}}$ can provide a high initial "boost" current to quickly reach V_{th}, followed by a slower transition through the Miller plateau where dv/dt control is critical.

5.19.3 Multi-Level Gate Driving (Stepped Driving)

Multi-level gate driving replaces the standard square-wave gate signal with a stepped voltage profile. Instead of a single transition from $V_{\text{GS,off}}$ to $V_{\text{GS,on}}$, the driver introduces one or more intermediate voltage levels.

Turn-On Sequence By dwelling briefly at an intermediate voltage (V_{int}), the driver reduces the di/dt during the initial phase of turn-on. This "soft start" minimizes the voltage overshoot on the companion diode and reduces high-frequency EMI harmonics.

Turn-Off Sequence Stepping the voltage down to an intermediate level before the final pull-down to negative bias helps "quench" the drain-source voltage rise, effectively acting as an active snubber without the efficiency penalty of passive components.

5.19.4 Active Gate Driving (AGD)

Active Gate Driving (AGD) represents the state-of-the-art in SiC control, where the gate drive current is adjusted in real-time based on closed-loop feedback.

- di/dt **Feedback:** A small sense inductor or a Rogowski coil in the source path provides a voltage proportional to the current slope. This signal is used to throttle the gate current if di/dt exceeds a safe limit.
- dv/dt **Control:** By sensing the displacement current through a small capacitor connected to the drain, the driver can adjust its strength to maintain a constant, pre-defined voltage slew rate.
- **Miller Plateau Detection:** The driver monitors V_{GS} to detect the exact start and end of the Miller plateau, boosting the gate current *before* and *after* the plateau to reduce delay while restricting it *during* the plateau to control dv/dt.

5.19.5 Protection Timing Coordination

The final layer of advanced control is the coordination of protection events. A "smart" gate driver must handle conflicting signals; for example, if an Active Clamp is trying to turn the device *on* to save it from overvoltage, while the DESAT circuit is trying to turn it *off* due to a short circuit.

A typical high-performance coordination sequence follows:

1. **Blanking:** The driver ignores V_{DS} noise for ~ 250 ns after turn-on.
2. **Detection:** The DESAT comparator triggers as current spikes.
3. **Soft Turn-Off (STO):** The driver ignores the PWM signal and engages the high-impedance STO path to slowly ramp current down.

4. **Active Clamping:** If the STO is too slow and a voltage spike occurs, the active clamp overrides the STO to "bleed" the voltage safely.

5. **Latching:** The driver latches into a safe "Off" state and signals a system-level fault.

5.20 Current-Source Gate Drivers (CSGD)

Traditional gate drivers operate as voltage sources with a series resistance (V_{DRV}/R_G). While simple, this approach results in an exponential gate-current decay that is highly sensitive to temperature-induced drifts in V_{th}. By contrast, **Current-Source Gate Drivers (CSGD)** enforce a regulated current:

$$i_G(t) = I_{\mathrm{const}} \quad \text{or} \quad i_G(t) = f(\text{target profile})$$

regardless of the instantaneous gate-source voltage, provided the driver remains within its compliance voltage limits.

The primary advantage of CSGD in SiC systems is the **linearization of the Miller plateau**. Because the charge Q_{gd} must be removed to facilitate the V_{DS} transition, a constant current ensures a constant dv/dt, drastically simplifying EMI filter design. Furthermore, CSGD is inherently better for paralleled devices; by injecting an identical charge into each gate, the driver ensures synchronized switching even if individual threshold voltages vary. The main challenges remain the high-bandwidth feedback required to prevent oscillations with the gate-loop inductance and the increased silicon area required for the constant-current regulators.

5.21 dv/dt and di/dt Shaping

The native switching speed of SiC MOSFETs, often exceeding 100 V/ns, is a double-edged sword. While it minimizes switching energy (E_{sw}), it subjects motor windings to partial discharge stress and generates significant common-mode EMI. **Switching-shaping** is the process of tailoring the gate current profile specifically during the "critical intervals" of the transition.

Advanced shaping techniques include:

- **Digital Resistor Banks:** Dynamically switching between different R_G values during a single transition to "soften" the start and end of the dv/dt slope.

- **Active Miller Clamping with Adaptive Strength:** Varying the clamping impedance based on the measured dv/dt of the complementary switch.

- **Hybrid AGD:** Using a high-current "boost" to reach the Miller plateau quickly, followed by a low-current "shape" phase to limit the slew rate.

5.22 Digital Gate Drivers and Adaptive Control

Digital gate drivers represent the integration of high-speed logic (FPGAs or DSPs) with the isolated power stage. Unlike analog drivers, a digital driver can "perceive" the switching environment and adapt in real-time. This enables **Prognostics and Health Monitoring (PHM)**, where the driver monitors shifts in the Miller plateau duration or V_{th} to predict device end-of-life.

Key capabilities include:

- **Adaptive Dead-Time:** Measuring the actual V_{GS} fall time to minimize dead-time, thereby reducing body-diode conduction losses.

- **Ageing Compensation:** Automatically decreasing R_G over the inverter's life to compensate for the gradual increase in V_{th} caused by interface trapping.

- **Fault Logging:** Capturing high-resolution waveforms of the microsecond preceding a DESAT event for post-mortem failure analysis.

5.23 AI-Driven Adaptive Gate Driving

In mission-critical applications, machine-learning (ML) algorithms can be deployed to optimize the "cost function" of a converter—balancing efficiency, EMI, and reliability.

5.23.1 Control Variables and Feedback Signals

The AI controller typically manipulates $R_{g,\text{on}}$, $R_{g,\text{off}}$, and the intermediate voltage levels of a multi-level driver. Inputs are derived from high-speed sensing of dv/dt, ringing frequency, and junction temperature (T_j). For example, a Reinforcement Learning (RL) agent might "experiment" with slightly different gate profiles during low-load conditions to find the optimal balance that minimizes V_{DS} overshoot without exceeding EMI limits.

5.23.2 Safety Envelopes

A critical requirement for AI-driven control is the **Safety Envelope**. The ML agent is strictly bounded by hardware-level comparators. If the AI suggests a gate profile that results in a V_{DS} spike near the breakdown voltage, the hardware "override" immediately reverts the driver to a conservative, pre-certified fallback setting.

5.24 Parallel MOSFET Management and Current Balancing

Paralleling SiC MOSFETs is necessary for high-power traction inverters but is notoriously difficult due to the "current crowding" effect. Even a 100 mV mismatch in V_{th} can lead to a 20% imbalance in transient current.

To ensure symmetry, the driver must manage:

- **Electrical Symmetry:** Using identical PCB trace lengths for the gate and Kelvin source of every die in the module.
- **Active Balancing:** Implementing individual, digitally-tuned gate delays for each parallel switch to align the start of the Miller plateaus.
- **Thermal Sensitivity:** While the positive temperature coefficient of SiC $R_{\text{DS,on}}$ helps balance *static* current, the *transient* current is dominated by V_{th}, which has a negative temperature coefficient, potentially leading to a localized thermal runaway if not corrected by the driver.

5.25 Active Miller Clamp

The high dv/dt of the "on-going" switch in a half-bridge induces a displacement current through the C_{gd} of the "off-going" switch ($i_{\text{Miller}} = C_{\text{gd}} \cdot dv/dt$). If this current flows through the gate resistor, it can lift V_{GS} above V_{th}, causing a catastrophic shoot-through.

The **Active Miller Clamp** mitigates this by providing a nearly zero-impedance bypass path. Once the gate driver senses that V_{GS} has dropped below a safety threshold (e.g., 2 V), an internal MOSFET shorts the gate directly to the source (or negative supply). This "locks" the device in the off-state, allowing for the use of smaller (or even zero) negative bias voltages in some cost-sensitive designs.

5.26 Gate-Energy Shaping for EMI Reduction

The total gate energy delivered during a cycle is defined as:

$$E_G = \int_{t_{\text{on}}}^{t_{\text{off}}} v_G(t) \cdot i_G(t)\, dt$$

By controlling the current *trajectory* rather than just the peak voltage, designers can "smear" the EMI spectrum. Instead of a single sharp peak in the MHz range, energy-shaping allows the driver to create multiple smaller, lower-amplitude harmonic peaks. This is particularly effective in aerospace systems where the weight of passive EMI filters is a primary design constraint.

5.27 Summary of Advanced Gate Control

The advanced gate-drive techniques that extend the performance and reliability of SiC MOSFETs beyond traditional fixed-resistor methods can be summarized as:

- Active gate shaping allows precise control of switching transitions and dv/dt.

- Multi-level gate driving balances speed, EMI, and robustness without sacrificing conduction performance.

- Current-source gate drivers enable consistent behavior across devices and operating conditions and enhance parallel-current sharing.
- Digital gate drivers unlock adaptive control, per-device calibration, and integration with system diagnostics.
- Parallel MOSFETs require advanced balancing and individual gate optimization to avoid current hogging and local overheating.
- EMI and reliability improve significantly when the switching trajectory and gate energy are intentionally shaped.

These methods must be supported by appropriate layout and isolation practices, discussed next.

5.28 Layout, EMI, and Practical Gate-Drive Implementation

The exceptional performance of SiC MOSFETs—characterized by nanosecond switching and minimal energy loss—is fundamentally limited by the physical environment of the power stage. In high-speed switching applications, the gate driver, PCB layout, and module packaging must be treated as a single, distributed electromagnetic system. A layout that is sufficient for Silicon IGBTs will often fail in a SiC context, leading to:

- **Voltage Overshoot:** Excessive V_{DS} spikes that threaten the device's breakdown voltage (BV_{DSS}).
- **Parasitic Turn-on:** Miller-induced gate transitions that cause catastrophic shoot-through.
- **EMI Compliance Failures:** High-frequency noise that exceeds CISPR or MIL-STD limits.
- **Protection Instability:** Nuisance tripping of DESAT or OCP circuits due to common-mode noise injection.

This section details the design constraints and layout strategies required to translate SiC semiconductor potential into system-level reliability.

5.29 Gate-Loop Inductance Minimization

The gate-loop inductance, $L_{\text{G-loop}}$, is the primary bottleneck for switching speed. It dictates the damping ratio of the gate signal and introduces propagation delays. Any voltage induced in this loop directly opposes the driver's command:

$$V_L = L_{\text{G-loop}} \frac{di_g}{dt}.$$

With SiC drivers often providing peak gate currents (i_g) of 5–10 A with rise times under 10 ns, even a few nanohenries of inductance can produce several volts of transient noise, causing the gate signal to ring or "stutter" during the Miller plateau.

5.29.1 Design Guidelines

To suppress $L_{\text{G-loop}}$, the gate and source return (Kelvin) traces must be kept in the closest possible proximity. Ideally, these should be routed on adjacent PCB layers to facilitate **magnetic flux cancellation**. Designers should avoid the use of stubs or long traces between the driver IC and the MOSFET; the distance should be minimized to the absolute physical limit of the package clearance.

5.30 Common-Source Inductance (CSI) and Feedback Effects

Common-Source Inductance (L_{CS}) is the parasitic inductance shared by both the gate-drive loop and the high-current power loop. As the drain current (i_D) rises rapidly during turn-on, the voltage drop across L_{CS} effectively subtracts from the applied gate voltage:

$$V_{\text{GS,eff}} = V_{\text{GS}} - L_{\text{CS}} \frac{di_D}{dt}.$$

This negative feedback mechanism slows the switching transition, negating the low-loss benefits of SiC. For high-current modules where di/dt can exceed 5 kA/μs, a mere 1 nH of CSI can reduce the effective gate-source voltage by 5 V, potentially keeping the device in the linear region for longer than intended. The use of **Kelvin-Source** pins is non-negotiable in high-performance designs, as they provide a dedicated, low-current return path for the gate driver that is isolated from the main power-loop voltage drops.

5.31 Power-Loop Layout and Flux Cancellation

The power loop—the path from the DC-link capacitors through the MOSFETs—is the largest source of radiated EMI and voltage overshoot. The parasitic inductance of this loop ($L_{\text{P-loop}}$) must typically be kept below 10–15 nH to prevent destructive ringing at turn-off:

$$\Delta V = L_{\text{P-loop}} \frac{di}{dt}.$$

5.31.1 Laminated Busbars and Vertical Stacking

In high-power systems, laminated busbars provide a low-inductance path by "sandwiching" positive and negative rails with thin dielectric layers. On PCBs, this is replicated by stacking the DC+ and DC- planes vertically on internal layers. This orientation forces the magnetic fields generated by the opposing currents to cancel each other out, drastically reducing the total stored energy in the parasitic magnetic field.

5.32 EMI Mechanisms and Common-Mode Transient Immunity

SiC MOSFETs generate significant Electromagnetic Interference (EMI) due to dv/dt levels often exceeding 100 V/ns. This high slew rate drives displacement currents through parasitic capacitances ($i = C \cdot dv/dt$) to the chassis or heatsink, manifesting as **Common-Mode (CM) noise**.

5.32.1 Isolation and CMTI

Gate drivers must provide galvanic isolation with high Common-Mode Transient Immunity (CMTI). A CMTI rating of > 100 kV/μs is standard for SiC applications. If the CMTI is insufficient, the rapid dv/dt of the power stage can couple through the isolation barrier's parasitic capacitance, causing the primary-side controller to latch up or the driver to issue false PWM signals. Modern capacitive or magnetic isolators are generally preferred over optocouplers for their superior CMTI and lower propagation delay skew.

5.33 Gate-Driver Power Supply Integrity

The gate-driver's isolated power supply must handle the same high dv/dt as the driver itself. A critical, often overlooked parameter is the **isolation capacitance** (C_{iso}) of the DC-DC converter. A large C_{iso} provides a low-impedance path for CM noise to flow back into the low-voltage control circuitry. Designers should select power modules with $C_{\text{iso}} < 10$ pF. Furthermore, negative gate bias rails (typically -3 to -5 V) must be tightly regulated; any sag in the negative rail during high-frequency switching reduces the safety margin against Miller-induced turn-on.

5.34 Digital Coordination and Prognostics

Modern SiC integration extends beyond simple pulse amplification. Advanced gate drivers now interact with FPGAs or DSPs to perform:

- **Adaptive Dead-Time:** Adjusting the delay between complementary switches in real-time based on measured turn-off speeds.
- **State-of-Health (SOH) Monitoring:** Using desaturation timing or gate-charge profiles to detect ageing indicators like bond-wire fatigue or gate-oxide degradation.
- **Active dv/dt Control:** Dynamically switching gate resistors (R_G) to meet EMI targets at low loads while maximizing efficiency at high loads.

5.35 Conclusion

This chapter has provided a comprehensive treatment of gate-drive design, protection, and control for SiC MOSFETs, building directly on the device physics (Chapters 2–3) and packaging concepts (Chapter 4). The main conclusions are:

- SiC gate-drive design must tightly control gate voltages, dv/dt, and di/dt to balance efficiency, EMI, and reliability.

- Protection circuits—especially DESAT detection with soft turn-off and negative gate bias—are indispensable because of SiC's limited short-circuit withstand time.
- Advanced gate-control techniques, including multi-level driving, AGD, and CSGD, enable tailored switching trajectories and better management of EMI and parallel devices.
- Layout and isolation practices are inseparable from gate-drive performance; inductance, CMTI, and parasitic coupling must be managed at the PCB and module level.
- Digital integration allows adaptive, intelligent gate control that responds to mission profiles, ageing, and changing operating conditions.

Together, these concepts define the gate-driver as a critical, system-level enabler for SiC power electronics.

Chapter 6

Future Trends in SiC Power Electronics

6.1 Introduction

Silicon Carbide (SiC) power technology has progressed from a specialized wide-bandgap option into a mainstream enabler of high-efficiency, high-density power conversion across transportation, renewable energy, aerospace, industrial automation, and emerging medium-/high-voltage (MV/HV) infrastructure. The material advantages of SiC—high critical electric field, reduced switching loss, high-temperature capability, and improved thermal conductivity—translate into converter-level benefits only when system constraints such as EMI, insulation coordination, packaging parasitics, thermal cycling, and fault protection are designed coherently.

Across the technical literature, a consistent pattern emerges: **the value of SiC increases with system voltage, switching frequency, and thermal/volume constraints**. High-voltage and high-density platforms (e.g., 800–1000 V EV powertrains, 1500 V PV strings, MV drives, HVDC submodules, and aircraft electrification) exhibit the clearest system-level benefits and the strongest economic pull toward SiC adoption.

This chapter surveys major application sectors and synthesizes the technical constraints that govern adoption. It also consolidates future trends in device structures, packaging, gate driving, sensing/health monitoring, and MV/HV system architectures that are shaping the next decade of SiC development.

6.2 Applications Driving SiC Adoption

This chapter surveys *deployed* application domains where SiC is already shaping commercial architectures, and it emphasizes the dominant stressors that link application requirements to device and packaging reliability. The goal is not to re-introduce SiC benefits in isolation, but to connect each application to its characteristic mission profile, insulation/EMI environment, thermal cycling regime, and protection requirements.

Emerging and frontier architectures that *change* the system boundary conditions (e.g., solid-state transformers, hydrogen infrastructure, and AI-enabled conversion) are treated separately in next chapters.

6.2.1 Electric Vehicles

Electric vehicles are the dominant near-term volume driver for SiC, particularly for 800 V-class architectures where switching loss reduction and thermal headroom deliver range and packaging benefits. The adoption trend is most visible in traction inverters, onboard chargers (OBCs), and high-voltage DC–DC stages.

Traction Inverters

SiC traction inverters enable higher switching frequency and reduced loss, but system limits are frequently set by EMI, insulation stress, and packaging-induced overshoot rather than by the intrinsic device speed. Practical performance is therefore tied to gate-driving strategy, parasitic control, and thermal stability. High-speed drive contexts (high electrical frequency machines) further reward SiC-based high-fundamental-frequency inverter architectures, and adaptive switching-frequency selection can improve system efficiency when magnetic loss and speed-dependent constraints are included.

Onboard Chargers and High-Voltage DC–DC

OBC and DC–DC stages increasingly leverage resonant or soft-switching structures, where switching speed and output capacitance influence circulating energy, light-load behavior, and EMI signatures. CLLC synchronous rectification strategies and resonant control observables (e.g., resonant capacitor voltage) are used to stabilize efficiency across load conditions. At higher power, resonant switched-capacitor conversion has emerged as a path to very high power density at high voltage in heavy transportation and aircraft-adjacent architectures.

6.2.2 Renewable Energy and Grid-Tied Power Conversion

PV inverters and energy storage systems (ESS) adopt SiC primarily for improved efficiency at high DC-link voltage and reduced cooling burden. Grid-tied systems also place strict requirements on power quality, thermal cycling, and long-life reliability. Active thermal control at the converter level can extend apparent kVA capability under grid constraints by shaping loss distribution and thermal headroom. At the system level, reliability and cost tradeoffs become central in multi-level inverter choices.

6.2.3 Aerospace and More-Electric Aircraft

Aerospace applications amplify constraints: extreme weight sensitivity, reduced convective heat transfer at altitude, aggressive reliability requirements, and stringent EMI/EMC. SiC enables lighter powertrains and higher-temperature operation, including cryogenic power electronics concepts that shift the thermal design space and component constraints. Demonstrations of SiC propulsion drives and flight validation highlight that success depends on packaging, thermal design, EMI control, and robust fault protection—not only device choice.

6.2.4 Industrial Automation and Motor Drives

Industrial drives adopt SiC for efficiency and dynamic response. System adoption often hinges on maintaining EMI compatibility with sensors, communications, and safety circuits. Converter-level switching behavior and commutation method selection affect both loss and stress, including comparisons of SiC MOSFET and Si IGBT solutions in four-quadrant switch behavior. Solid-state circuit breaker (SSCB) integration into motor control centers represents an architectural shift enabled by SiC, expanding protection and controllability at the distribution level.

6.2.5 Medium-Voltage and High-Voltage Systems (MV Drives and HVDC)

SiC is rapidly moving into MV domains traditionally dominated by silicon IGBTs/IGCTs. MV SiC system penetration depends on series connection, insulation coordination, gate-driving isolation, and suppression of gate oscillations and voltage imbalance. MV overcurrent protection and sensing architectures (e.g., Rogowski coil broadband sensors) are increasingly relevant at these scales. MV grid/traction converters are also exploring hybrid and multilevel approaches at high switching frequency.

6.3 Future Trends in SiC Device Technology and Ruggedness

SiC device evolution is increasingly defined by three coupled goals: (i) reduced $R_{\mathrm{DS(on)}}$ and switching loss, (ii) improved ruggedness and short-circuit tolerance, and (iii) predictable reliability under mission-profile stress, particularly for trench structures and high-voltage classes.

6.3.1 Trench MOSFET Roadmap and Reliability Engineering

Trench SiC MOSFETs deliver improved conduction performance but place higher demands on oxide field management, trap control, and robustness. Reliability engineering for advanced VMOSFET/trench technologies emphasizes electric-field shaping, oxide stability, and parameter drift suppression. Gate-related trap characterization methods and stress-leakage-based observables support deeper understanding of oxide/interface behavior in 4H-SiC MOSFETs. Degradation modeling under gate switching instability and parameter drift provides a pathway to design margins and screening strategies.

Device C–V modeling and temperature scaling remain important because device capacitances drive both switching behavior and EMI. Accurate capacitance extraction methods for EMI analysis, including package-aware half-bridge extraction of dynamic capacitances, enable more predictive converter models. Temperature-dependent SPICE and analytical models that incorporate parasitic parameters improve converter-level prediction across thermal operating ranges.

6.3.2 Third-Quadrant Operation, Surge, and Avalanche Robustness

As converters push higher switching frequency and exploit synchronous conduction, third-quadrant behavior becomes a system limiter. Investigations of surge robustness under various V_{GS} biases and planar-gate third-quadrant surge characteristics highlight that diode/body-diode behavior is still a key adoption constraint in some hard-commutation conditions. Body-diode degradation mechanisms linked to implantation doping under third-quadrant conduction further emphasize the need for structure-aware reliability screening. New device structures integrating Schottky contacts or barrier rectification aim to reduce third-quadrant loss and mitigate degradation paths. Avalanche UIS ruggedness comparisons also remain relevant for fault and transient robustness in certain systems.

6.3.3 Short-Circuit and Fault Ruggedness as a System Gate

Short-circuit withstand time and protection speed are central to deployment, especially for dense trench devices and MV applications. Temperature-dependent short-circuit protection schemes and rapid detection strategies (e.g., drain-voltage swing methods) address the reduced protection window. Combined V_{GS} and V_{DS} detection methods expand fault coverage and enable more robust protection logic. In MV devices, loss reduction strategies that do not increase dv/dt while maintaining DESAT protection become especially important.

6.3.4 Radiation and Space/Altitude Effects

Aerospace and space-adjacent deployments require understanding radiation effects and temperature-dependent leakage behavior. Proton-induced displacement damage and total-ionizing-dose effects in vertical and trench SiC MOSFETs, along with heavy-ion-induced leakage and single-event burnout sensitivity, define qualification constraints and derating policies. These effects are increasingly relevant as aircraft electrification expands power electronics into higher-altitude and higher-radiation mission envelopes.

6.4 Packaging, Gate Driving, EMI, and Intelligent Integration Trends

In modern SiC systems, packaging parasitics, driver behavior, and sensing/monitoring increasingly define achievable switching performance. The key trajectory is moving from discrete device-centric design toward integrated power stages with embedded drivers, sensing, and health monitoring.

6.4.1 Active Gate Driving, dv/dt Shaping, and Overshoot/Ringing Control

Active gate drivers are becoming a primary lever for suppressing overshoot and oscillation without sacrificing switching speed. Driver approaches include closed-loop dv/dt and di/dt regulation, independent overshoot suppression, and timing control integrated into driver ICs. Self-regulating overshoot suppression under variable load conditions and crosstalk mitigation in phase-leg configurations address practical half-bridge risks. Passive triggered clamping and level-shifter schemes further support gate control in constrained MV environments.

Because paralleled devices introduce current imbalance and gate-loop sensitivity, current sharing via auxiliary branches, delay/slew regulation, and Kelvin-source-aware balancing drivers are increasingly important. Screening-indicator selection and clustering methods for dynamic current balancing provide a systematic path to stable paralleling. Coupled inductor strategies for balancing and oscillation suppression complement driver-based approaches.

6.4.2 Series Connection for MV/HV SiC

To achieve Medium Voltage (MV) and High Voltage (HV) ratings using existing Silicon Carbide technology, designers often resort to series-connected MOSFET stacks. This approach bypasses the yield and availability issues of monolithic 10 kV+ devices but introduces significant complexities regarding voltage equalization, isolation, and synchronization.

Series-connected stacks require a multi-layered control strategy to prevent any single device from exceeding its breakdown voltage (BV_{DSS}) during switching transients or steady-state off-intervals.

Voltage Balancing Challenges:

Voltage imbalance in a series string typically arises from mismatches in device parameters—specifically the output capacitance (C_{oss}), threshold voltage (V_{th}), and leakage current—as well as asymmetries in the gate-drive signal timing.

- **Static Balancing:** Resistor networks are typically used to equalize the voltage across the stack during the off-state, compensating for variations in leakage current.
- **Dynamic Balancing:** During high-dv/dt switching events, active balancing is required. This often involves **active voltage clamping** or comparison circuitry that adjusts the gate-drive timing in real-time. By slightly delaying the turn-on or accelerating the turn-off of specific devices, the driver ensures that the V_{DS} slope is uniform across the entire string.
- **Body-Diode Turn-off:** In converter topologies like the Series Resonant Dual Active Bridge (SRDAB), the reverse recovery behavior of the body diode must be managed. Active balancing strategies are frequently employed here to prevent localized voltage spikes during the diode's commutation phase.

Gate-Drive Architectures for Stacks:

Scalability is a primary concern for MV stacks. Traditional isolated drivers require individual, high-isolation-voltage power supplies for every device in the string, which increases system volume and cost.

Isolated vs. Non-isolated Trends: While robust synchronization across a fiber-optic or capacitive isolation barrier is the industry standard, there is a parallel trend toward **single-gate or non-isolated supply architectures**. These "daisy-chain" or bootstrap-based methods aim to derive gate power from the main HV bus or through high-frequency magnetic coupling, significantly reducing the complexity of the auxiliary power system.

Stability and Gate Oscillation at 10 kV:

At ultra-high voltages and high chip counts (e.g., 10 kV modules or large paralleled-series arrays), the system becomes highly susceptible to **gate oscillation**. This is often triggered by the interaction between the high dv/dt and the parasitic capacitance of the isolation barrier and the module casing.

Mitigation strategies for 10 kV devices highlight that stability is no longer just a component-level concern but a system-level design target. Techniques such as damping resistors in the gate path, localized active Miller clamps, and careful management of the common-mode current paths are essential to prevent oscillations from leading to catastrophic gate-oxide failure.

6.4.3 Packaging: Low Inductance, Thermal Density, and High-Temperature Encapsulation

Packaging increasingly determines achievable switching speed and thermal cycling lifetime. Trends include multichip module manufacturing innovations (e.g., direct ink writing), baseplateless module solutions, and symmetric/staggered terminals to reduce parasitics and improve current balance. Co-design frameworks for MV SiC modules emphasize that electrical inductance, thermal paths, and mechanical constraints must be optimized together at the module level. High-performance 3-D packaging structures and pressure-contact double-sided cooling modules target high-temperature and high power density operation.

High-temperature operation drives encapsulation and coating innovation, including hermetic conformal coating methods and thin-film encapsulation solutions aimed at maintaining integrity at 250°C-class operation. Integrated gate-driver modules operating at elevated temperature demonstrate the system-level path toward compact, rugged power stages. Warpage optimization in panel-level packaging highlights the growing importance of thermo-mechanical control as packaging scales.

6.4.4 Layout, Parasitic Extraction, and Conducted Common-Mode Current

As switching speeds increase, conducted common-mode current during transients is strongly influenced by layout constraints. Studies of module layout optimization for CM current reduction, and comprehensive comparisons of layout topologies, reinforce that parasitics (and the spatial distribution of capacitance/inductance) directly set EMI outcomes. Distributed decoupling capacitor placement can simultaneously suppress overshoot and improve dynamic current sharing. Power-loop stray inductance extraction methods (e.g., forced resonance with decoupling capacitors) support quantitative design closure on overshoot and ringing prediction. Voltage-driven reverse recovery modeling perspectives also support more accurate commutation and EMI prediction for SiC MOSFETs.

6.4.5 Thermal Modeling, Monitoring, and Health Prognostics

Thermal modeling and monitoring are shifting toward physics-informed and intelligent estimation methods that generalize across operating conditions. Physics-informed intelligent thermal models for SiC modules and electrical-thermal coupling approaches support digital-twin design flows. Integrated thermal management approaches at the module or converter level aim to suppress hotspots and improve thermal cycling lifetime. Electrothermal interaction in linear-mode operation across wide temperature ranges is increasingly relevant for abnormal states and protection events.

Junction temperature extraction and estimation remain central. Comprehensive reviews of T_j extraction, efficient estimation methods, and online temperature monitoring circuitry for WBG transistors provide practical implementation pathways. High-accuracy converter-level on-state voltage measurement, gate-voltage dip indicators for health monitoring, and delay-time based temperature monitoring over aging demonstrate how electrical observables are used for embedded health assessment. Self-sensing methods for diagnosing thermal-interface aging reinforce that packaging degradation is observable and monitorable in-field. Novel transient thermal characterization systems and pulse current sources for thermal impedance measurement support improved validation and parameter identification.

6.4.6 Power Cycling, Lifetime Modeling, and RUL Prediction

Power cycling methodologies and limitations—especially when switching loss magnitude influences lifetime during cycling—highlight the need for carefully defined test conditions and mission-profile correlation. Physics-informed deep learning methods for remaining useful life (RUL) prediction under sparse data reflect the trend toward intelligent prognostics as SiC enters safety-critical domains. Driver-integrated leakage detection and indirect bond-wire liftoff monitoring extend monitoring beyond temperature into packaging degradation signatures. Hyperbolic deep learning frameworks for intelligent gate driving and health monitoring further illustrate the convergence of control and reliability telemetry.

6.4.7 Converter Architectures Enabled by High-Frequency Operation

The superior switching characteristics of Silicon Carbide (SiC) MOSFETs—specifically the reduction in switching energy (E_{on}, E_{off}) and the nearly negligible reverse-recovery charge (Q_{rr})—have fundamentally shifted the design paradigms of power conversion. By moving switching frequencies into the hundreds of kHz or low MHz range, designers can implement architectures that were previously considered impractical due to thermal or efficiency constraints in Silicon (Si) based systems.

Wireless Power Transfer (WPT) and High-Q Systems:

High-frequency (HF) operation is particularly transformative for Wireless Power Transfer (WPT). Operating at the MHz frequencies required by standards such as SAE J2954 necessitates low gate charge (Q_g) and output capacitance (C_{oss}) to maintain high efficiency. SiC-enabled inverters allow for an optimized parallel chip count under sophisticated pulse-width control, ensuring that the inverter maintains high-quality factor (Q) resonance even under varying coupling coefficients and load conditions. This frequency agility enables the use of smaller, lighter magnetic resonators, which is a critical requirement for electric vehicle (EV) charging and aerospace applications.

Resonant and Soft-Switching Topologies:

Beyond traditional hard-switched converters, SiC switching behavior has revitalized interest in specialized resonant-pole and auxiliary commutated structures.

- **Soft-Switching Synergies:** While SiC reduces hard-switching losses, its integration into Zero-Voltage Switching (ZVS) and Zero-

Current Switching (ZCS) topologies—such as LLC and CLLC resonant converters—enables ultra-high power density. The low $R_{\text{DS,on}}$ minimizes conduction losses, while the fast output capacitance transition facilitates rapid ZVS commutation, allowing for smaller heatsinks and passive components.

- **Harmonic Performance:** High-frequency operation naturally shifts the harmonic content to higher orders, simplifying the design of output filters and achieving exceptionally low Total Harmonic Distortion (THD) in grid-tied inverters and motor drives.

Solid-State Circuit Breakers (SSCB) and Fast Protection:

The transition toward SiC also facilitates a move toward reconfigurable and "intelligent" distribution networks through **Solid-State Circuit Breakers (SSCB)**. Unlike mechanical breakers that operate on millisecond timescales, SiC SSCBs can interrupt fault currents in microseconds.

- **Advanced Snubber Concepts:** To manage the extreme di/dt during a fault interruption, advanced snubber concepts are employed. This includes **inverse current injection** techniques, where a counter-current is injected to force a zero-crossing, allowing the SiC device to safely commutate the inductive energy of the line.
- **System-Level Reconfigurability:** The high-speed response of SiC allows for a more granular approach to fault isolation in DC microgrids and MVDC (Medium Voltage DC) systems, enabling "fault-ride-through" capabilities that were previously unattainable.

Ultimately, these high-frequency architectures represent a broader shift from "bulk" power conversion to high-density, software-defined power systems where the semiconductor's switching speed is the primary lever for system-level optimization.

6.5 Cross-Domain Synthesis, Roadmap, and Outlook

The trajectory of Silicon Carbide adoption is no longer dictated by device feasibility; the industry has moved past the question of "if" SiC works to the challenge of "system closure." True success in the next decade will be defined by how effectively we manage the coupled constraints of EMI, insulation stress, protection speed, thermo-mechanical packaging, and long-term lifetime predictability. The roadmap for SiC integration is converging around four pillars.

6.5.1 Theme 1: Adaptive Switching - From Maximum Velocity to Contextual Optimization

We are witnessing a departure from the "switch as fast as possible" philosophy. The industry is moving toward **adaptive switching behavior**, where dv/dt and di/dt are dynamically tuned based on instantaneous EMI margins, thermal headroom, and the estimated state-of-health of the device.

The enabling infrastructure for this shift includes closed-loop gate drivers, high-bandwidth timing ICs, and active overshoot suppression. By utilizing predictive modeling of crosstalk and temperature-dependent transients, next-generation controllers will treat switching speed as a programmable variable rather than a fixed hardware constraint, allowing converters to operate at peak efficiency during light loads while softening transitions to meet strict EMI targets during high-power bursts.

6.5.2 Theme 2: Interconnect and Packaging - The New Frontier of Power Density

As the semiconductor die shrinks, the package increasingly becomes the primary bottleneck for power density. Improvements in volumetric efficiency now come from 3-D module structures, double-sided cooling, and pressure-contact designs that eliminate the inductive penalties of traditional bond wires.

Furthermore, the shift toward sustained high-temperature operation ($> 175°C$) demands a revolution in material science. High-temperature encapsulation, advanced ceramic coatings, and thermo-mechanical warpage control are now essential to prevent insulation breakdown or mechanical fatigue. In this new era, the "power module" is evolving into a highly integrated "power substrate," where sensing, protection, and cooling are fused into a single manufacturable unit.

6.5.3 Theme 3: System-Level Reliability - Beyond the Semiconductor Die

Reliability is no longer a localized semiconductor metric; it is a coupled behavior involving gate-oxide traps, third-quadrant conduction fatigue, and packaging delamination. Characterizing gate-switching instabilities and surge robustness provides the data necessary for realistic derating and screening.

Qualification standards are also evolving to align with specific **mission profiles**. Rather than generic power cycling, developers now focus on how the magnitude of switching losses influences cycling lifetime in real-world drive cycles. The ultimate goal is the implementation of in-field health monitoring—using gate voltage signatures and self-sensing thermal-interface diagnostics to move from reactive maintenance to a predictive, fleet-wide reliability model.

6.5.4 Theme 4: MV/HV Infrastructure - Scaling the SiC Advantage

For SiC to dominate the Medium Voltage (MV) and High Voltage (HV) sectors, it must overcome the challenges of series-connected scaling and scalable isolation. This requires an infrastructure of active voltage balancing, high-isolation gate-driver supplies, and broadband sensing tools such as Rogowski coils for nanosecond fault detection.

In the 10 kV class, mitigating gate oscillation is a core design target, as the high dv/dt can easily excite parasitic resonances in large-scale arrays. While emerging materials like Ga_2O_3 (Gallium Oxide) offer theoretical performance gains, SiC's leadership in the MV/HV space will be maintained through its superior ruggedness, established reliability data, and the availability of a scalable series-protection infrastructure.

6.5.5 Outlook by Time Horizon

The evolution of Silicon Carbide is not a monolithic leap but a series of phased transitions defined by voltage classes, manufacturing maturity, and the readiness of the surrounding ecosystem. As we stand in 2026, the roadmap is clearly bifurcated into immediate industrial scaling and long-term infrastructure transformation.

Near-Term (2026–2027): EV Dominance and the Rise of Intelligent Power Modules (IPM)

In the immediate horizon, Silicon Carbide is synonymous with the global transition to 800 V Electric Vehicle (EV) platforms. The focus has shifted from device-level performance to **integrated system closure**.

- **High-Temperature Integration:** We are seeing the mass production of 200°C-class phase-leg modules that integrate the gate driver, desaturation protection, and temperature sensing into a single, hermetically sealed unit. This reduces the parasitic distance between the driver and the die to sub-millimeter scales, effectively "designing out" gate-loop inductance.

- **Manufacturing Maturity:** The industry-wide transition from 150 mm to 200 mm (8-inch) wafers is now a factory-floor reality rather than a pilot-line promise. This scale-up is driving the cost-per-die down, making SiC competitive even in cost-sensitive industrial motor drives and high-end consumer power supplies.

Mid-Term (2027–2031): Medium-Voltage (MV) Expansion and Architecture Evolution

By the end of the decade, the center of gravity for SiC will shift toward Medium Voltage (MV) applications (3.3 kV to 10 kV). This era will be defined by the maturation of **series-stacked architectures** and hybrid multilevel converters.

- **Active Balancing Standardization:** The complex active voltage balancing and synchronization strategies discussed in earlier chapters will move from research prototypes into standardized controller libraries. This allows for modular MV converters that are as easy to deploy as today's low-voltage inverters.

- **Resonant and Switched-Capacitor Innovation:** High-frequency switching will enable a "magnetics-free" or "magnetics-light" revolution in high-power domains. Topologies like the CLLC and dual-active bridge (DAB) will become the standard for bi-directional DC-DC conversion in grid-tied storage and aerospace microgrids.

Long-Term (Beyond 2031): Grid Infrastructure and Self-Healing Power Stages

In the long term, Silicon Carbide will be the primary lever for the modernization of the utility grid. As 15 kV+ devices reach commercial maturity, we will see the widespread penetration of Solid-State Transformers (SST) and HVDC/MMC (Modular Multilevel Converter) stations.

- **Highly Integrated Power Stages:** The power stage of 2031 will be "self-sensing." Intelligent gate-driving frameworks will be standard, providing continuous, AI-driven reliability prediction and autonomous fault-ride-through capabilities.
- **Insulation Coordination:** Practical experience from the previous decade will result in a set of rigorous insulation coordination practices, specifically regarding the high-dv/dt stress on motor windings and grid transformers. SiC systems will not only be fast and efficient but will possess the "utility-grade" ruggedness currently associated with legacy Silicon Thyristors.

Ultimately, the goal is a fully software-defined power stage, where Silicon Carbide provides the high-speed muscle and digital coordination provides the intelligence.

6.6 Conclusion

SiC is reshaping power electronics by enabling new combinations of efficiency, switching frequency, temperature capability, and power density. The next decade of SiC progress will be defined less by raw device speed and more by **system closure**: scalable gate driving (including MV stacking), low-inductance high-thermal-density packaging, EMI/insulation coordination, fast and reliable protection, and embedded sensing/health monitoring that turns reliability into a controllable system variable.

Bibliography

[1] B. J. Baliga. “Semiconductors for high-voltage, vertical channel field-effect transistors”. In: *Journal of Applied Physics* (1982).

[2] T. Kimoto and J. A. Cooper. *Fundamentals of Silicon Carbide Technology*. Wiley-IEEE Press, 2014.

[3] H. Matsunami. “Fundamental research on semiconductor SiC and its applications to power electronics”. In: *Proceedings of the Japan Academy, Series B* (2004).

[4] Wolfspeed. “The Transition from 150mm to 200mm SiC Wafers: Yield and Cost Implications”. In: *International Conference on Silicon Carbide and Related Materials (ICSCRM)*. 2022.

[5] P. G. Neudeck, R. S. Okojie, and L. Y. Chen. “High-temperature electronics - a role for wide bandgap semiconductors?” In: *Proceedings of the IEEE* (2002).

[6] Yole Group. *Power SiC 2024: Market and Technology Report*. Tech. rep. Yole Intelligence, 2024.

[7] S. T. Pantelides et al. “SiC/SiO2 Interface: Physics and Reliability”. In: *Materials Science Forum* (2006).

[8] C. Lombardi et al. “A physically based mobility model for numerical device simulation”. In: *IEEE Transactions on Computer-Aided Design of Integrated Circuits and Systems* (1988).

[9] G. Y. Chung et al. “Improved Inversion Channel Mobility for 4H-SiC MOSFETs Following High Temperature Anneals in Nitric Oxide”. In: *IEEE Electron Device Letters* (2001).

[10] D. Peters et al. “The CoolSiC MOSFET - Design concept and performance update”. In: *PCIM Europe*. 2017.

[11] T. Nakamura. "Latest 4th Generation SiC MOSFET Technology for Automotive Applications". In: *APEC*. 2020.

[12] A. J. Lelis et al. "Basic Mechanisms of Threshold-Voltage Instability in SiC MOSFETs". In: *IEEE Transactions on Electron Devices* (2009).

[13] A. Agarwal. "SiC MOSFET Reliability: Current Status and Future Challenges". In: *IEEE Transactions on Device and Materials Reliability* (2017).

[14] L. Zhang et al. "TDDB Reliability Comparison of Planar and Trench SiC MOSFETs". In: *IEEE Journal of Emerging and Selected Topics in Power Electronics* (2020).

[15] T. Aichinger et al. "Threshold voltage drift of SiC MOSFETs under static and dynamic bias stress". In: *Microelectronics Reliability* (2018).

[16] M. Skowronski and S. Ha. "Degradation of SiC bipolar devices". In: *Journal of Applied Physics* (2006).

[17] P. Bergman et al. "Crystal defect problems in SiC bipolar devices". In: *Materials Science Forum* (2001).

[18] T. Mizoguchi et al. "Improved reliability of 1.2kV SiC MOSFET by preventing the intrinsic body diode operation". In: *PCIM Europe*. 2020.

[19] P. Reigosa et al. "Short-Circuit Failure Mechanism of 1200 V SiC MOSFETs". In: *IEEE Transactions on Industrial Electronics* (2017).

[20] F. Iannuzzo et al. "A Review of SiC MOSFET Short-Circuit Ruggedness and Failure Mechanisms". In: *IEEE Journal of Emerging and Selected Topics in Power Electronics* (2015).

[21] P. Friedrichs. "SiC MOSFETs: From Device Physics to System Reliability". In: *IEEE Power Electronics Magazine* (2021).

[22] H. Schwarzbauer and R. Beck. "Novel joining technique for power electronic devices". In: *IEEE Transactions on Components, Hybrids, and Manufacturing Technology* (1991).

[23] E. Hoene et al. "Ultra low inductance package for SiC". In: *PCIM Europe*. 2007.

[24] M. Kozak et al. "Thermal modeling of SiC power modules using Cauer and Foster networks". In: *IEEE Transactions on Industry Applications* (2015).

[25] M. Chen. "SiC Power Modules with Silver Sintered Molybdenum Packaging". McMaster University, 2025.

[26] M. Liserre et al. "Active Gate Driving Strategies for WBG Semiconductors". In: *IEEE Transactions on Power Electronics* (2023).

[27] H. Zhang et al. "A new active gate driver for improving the switching performance of SiC MOSFET". In: *IEEE Xplore* (2017).

[28] N. Baker et al. "Gate-loop inductance minimization in SiC power modules". In: *IEEE Journal of Emerging and Selected Topics in Power Electronics* (2019).

[29] A. Q. Huang. "Medium-Voltage Power Electronics: Powering the Future Grid". In: *Proceedings of the IEEE* (2017).

[30] J. Palmour et al. "10 kV - 15 kV Silicon Carbide Power MOSFETs for Next-Generation Energy Conversion". In: *IEEE ECCE*. 2014.

[31] IEEE Open Journal of Power Electronics. "Overview of Digital Twin Development in Power Electronics". In: (Sept. 2025).

[32] IEEE Conference Publication. "AI-Driven Simulation and Digital Twin Technologies for Smart Business Planning". In: (Feb. 2026).

[33] JEDEC Committee JC-70. *Wide Bandgap Power Electronic Conversion Semiconductors: SiC Standards*. Tech. rep. JEDEC, 2018.

[34] *High Temperature Reverse Bias (HTRB) and High Temperature Gate Bias (HTGB)*. JEDEC, 2021.

[35] S. Wang et al. "Impact of Cosmic Radiation on 3.3kV SiC MOSFETs in Rail Traction". In: *IEEE Transactions on Transportation Electrification* (2024).

[36] Y. Liu et al. "Health Monitoring of SiC MOSFETs using AI-based On-State Voltage Sensing". In: *IEEE Transactions on Industrial Informatics* (2025).

[37] Yu Xiao et al. "Partial Power Converter Based on the Gate Voltage Self-Balancing Supercascode Power Switch". In: *IEEE Transactions on Power Electronics* (2026).

[38] Ho-Tin Tang and Henry Shu-Hung Chung. "Switched-Capacitor-Based Gate Driver With Real-Time Gate–Source Leakage Current Detection and Indirect Bond Wire Liftoff Monitoring". In: *IEEE Transactions on Power Electronics* (2025).

[39] Yanchao Liu et al. "Current Balancing and Voltage Oscillation Suppression of Parallel SiC MOSFETs With Compact Coupled Inductors". In: *IEEE Journal of Emerging and Selected Topics in Power Electronics* (2025).

[40] Hu Tan et al. "Three-Level T-Type APF With Paralleled Discrete SiC MOSFETs Using Optimized-Balanced Parasitic Inductance Structure". In: *IEEE Journal of Emerging and Selected Topics in Power Electronics* (2025).

[41] Zizhao Wang et al. "A Novel Short-Circuit Protection Scheme for Silicon Carbide (SiC) MOSFET Module Considering Operation Temperature". In: *IEEE Transactions on Power Electronics* (2025).

[42] Cheng Li et al. "A Self-Regulating Active Gate Driver of Voltage Overshoot Suppression for SiC MOSFETs Under Variable Load Current Conditions". In: *IEEE Journal of Emerging and Selected Topics in Power Electronics* (2025).

[43] Peng Sun et al. "Design Method of Distributed Decoupling Capacitors for Both Voltage Overshoot Suppression and Dynamic Current Sharing in SiC MOSFET Power Module". In: *IEEE Transactions on Industrial Electronics* (2025).

[44] Ming Zhang et al. "RMS Current Balancing of Parallel SiC Bridge Arms Based on a Rotational Rest Control Strategy". In: *IEEE Transactions on Power Electronics* (2026).

[45] Li Zhang et al. "SiC MOSFET Hybrid Switches With Reverse Conducting IGBTs: Loss and Surge Current Analysis". In: *IEEE Transactions on Industrial Electronics* (2026).

[46] Yibo Gao et al. "A Hermetic Conformal Coating Based High-Temperature Encapsulation Method for 250 °C SiC Power Module". In: *IEEE Transactions on Power Electronics* (2025).

[47] Zeyu Liu et al. "One-Step Method of Dynamic Capacitances Extraction From a SiC Power MOSFET in a Half-Bridge Package for EMI Analysis". In: *IEEE Transactions on Power Electronics* (2025).

[48] Jun Wang et al. "Improved Model for Accurate Prediction of Crosstalk-Induced Gate–Source Voltage Peaks of SiC MOSFET With SiC Schottky Diode in a Half-Bridge Configuration". In: *IEEE Journal of Emerging and Selected Topics in Power Electronics* (2025).

[49] Ming Zhang et al. "Design and Validation of a 100 kHz Inner-Paralleled Si+SiC Hybrid Multilevel Converter for Medium-Voltage Electric Traction Drives". In: *IEEE Transactions on Transportation Electrification* (2026).

[50] Hong Chen et al. "Degradation Evaluation for Ion Implantation Doping in 1.2kV SiC MOSFET Body Diode Under Third Quadrant Conduction". In: *IEEE Transactions on Device and Materials Reliability* (2025).

[51] Haotian Zhang et al. "A Brief Review of SiC MOSFET Transient Analytical Modeling Methods: Principles, Current Status, and Parameters Modeling". In: *IEEE Transactions on Power Electronics* (2025).

[52] Cheng Li et al. "Active Gate Driver With Auxiliary Branch for Current Sharing of Paralleled SiC MOSFETs". In: *IEEE Journal of Emerging and Selected Topics in Power Electronics* (2026).

[53] Zizhao Wang et al. "Electrical-Thermal Coupling Modeling of SiC MOSFETs Based on Field-Circuit Coupling and Its Application in Junction Temperature Calculation During Surges". In: *IEEE Transactions on Industrial Electronics* (2025).

[54] Yibo Gao et al. "Development, Integration, and Flight Testing of a Silicon Carbide Propulsion Drive for a Hybrid Electric Aerospace Application". In: *IEEE Transactions on Transportation Electrification* (2026).

[55] Wei Chen et al. "Analysis and Mitigation Methods of Gate Oscillation in Paralleled 10 kV SiC MOSFETs". In: *2025 IEEE Applied Power Electronics Conference and Exposition (APEC)*. 2025.

[56] Yanchao Liu et al. "500-kW Resonant Switched-Capacitor Converter for 1600-V Electric Trucks and Electric Aircraft Powertrain Application". In: *IEEE Transactions on Transportation Electrification* (2026).

[57] Hu Tan et al. "Co-design Framework for High Power, Medium/High Voltage WBG Power Modules: Case Study with 3.3 kV/200 A Wire-Bonded Low-Inductance SiC Half-Bridge Module". In: *IEEE Open Journal of Power Electronics* (2025).

[58] Mingxue Huo et al. "Remaining Useful Life Prediction of Power Electronic Devices With Physics-Informed Deep Learning and Sparse Data". In: *IEEE Transactions on Power Electronics* (2025).

[59] Zeyu Liu et al. "Influence of Switching Loss Magnitude on Lifetime During a Switch-Mode Power Cycling Test of SiC MOSFETs". In: *IEEE Transactions on Device and Materials Reliability* (2025).

[60] Hong Chen et al. "Turn-on Switching Loss Reduction for 10 kV SiC MOSFETs Without Increasing dv/dt Using Desat Protection Circuits". In: *IEEE Transactions on Power Electronics* (2025).

[61] Chao Huang et al. "Universal Active Gate Driver IC With Closed-Loop Timing Control and Gate-Sensing Technique for Silicon Carbide Power Devices". In: *IEEE Journal of Solid-State Circuits* (2026).

[62] Jun Wang et al. "Turn-OFF Delay Time Based Online Junction Temperature Monitoring for SiC MOSFETs Over Aging". In: *IEEE Transactions on Power Electronics* (2025).

[63] Ming Zhang et al. "A Pulse Current Source for Transient Thermal Impedance Measurement with Low Nonthermal Switching Transient Effects". In: *IEEE Transactions on Instrumentation and Measurement* (2025).

[64] Fengtao Yang et al. "Electrothermal Interaction Mechanism for SiC Power MOSFET in Linear-Mode Operation From 25°C to 375°C: Theoretical Analysis, Experimental Verification and Packaging Effect". In: *IEEE Journal of Emerging and Selected Topics in Power Electronics* (2025).

[65] Peng Sun et al. "Symmetric and Staggered Terminal Layouts for Enhanced Current Balance and Reduced Parasitic Inductance in SiC Power Modules". In: *IEEE Transactions on Power Electronics* (2025).

[66] Hao Li et al. "Efficient Junction Temperature Estimation of SiC Power Modules Based on Temperature-Dependent Lumped Thermal Model". In: *IEEE Transactions on Industrial Electronics* (2024).

[67] Zeyu Liu et al. "Comparison and Investigation on the Static and Dynamic Performance for Medium-Voltage (3300 V-6500 V) Ga2O3 and SiC Planar Gate MOSFETs". In: *IEEE Transactions on Electron Devices* (2025).

[68] Che-Wei Chang and Dong Dong. "Enhanced di/dt-RC Sensing Structure for Accurate Dynamic Current Measurement in Paralleled SiC MOSFETs". In: *IEEE Transactions on Power Electronics* (2025).

[69] Jun Wang et al. "Physical Understanding on Short-Circuit Failure for Cascode GaN HEMTs". In: *IEEE Transactions on Device and Materials Reliability* (2025).

[70] Peng Sun et al. "The Influence and Optimization Constraints of SiC Power Module Layout on High-Frequency Conducted CM Current During Switching Transients". In: *IEEE Transactions on Electromagnetic Compatibility* (2025).

[71] Cheng Li et al. "High Accuracy and Wide Temperature Range Converter-level On-State Voltage Measurement With Localization Function for SiC MOSFETs". In: *IEEE Transactions on Industrial Electronics* (2026).

[72] H. Wen et al. "Junction Temperature Extraction for Silicon Carbide Power Devices: A Comprehensive Review". In: *IEEE Transactions on Power Electronics* (2025).

[73] Hu Tan et al. "A T-Shaped High Isolation Gate Driver Power Supply for Medium-Voltage SiC Half-Bridges". In: *IEEE Open Journal of Power Electronics* (2025).

[74] Yibo Gao et al. "A Temperature-Dependent Analytical Transient Model of SiC MOSFET in Half-Bridge Circuits". In: *Sensors* (2025).

[75] Zeyu Liu et al. "A Double-Sided Cooling SiC Power Module With Pressure Contact Packaging". In: *IEEE Transactions on Components, Packaging and Manufacturing Technology* (2025).

[76] Rong Zhang et al. "Thin-Film Encapsulation Solution for High-Temperature SiC Power Devices". In: *IEEE Transactions on Power Electronics* (2025).

[77] M. R. Nielsen et al. "High-Power Electronic Applications Enabled by Medium Voltage Silicon-Carbide Technology: An Overview". In: *IEEE Transactions on Power Electronics* (2024).

[78] Jun Wang et al. "An Online Unified Delay and Slew Rate Regulation for Current Sharing in Paralleled SiC Power Modules With Active Gate Drivers". In: *IEEE Transactions on Industrial Electronics* (2025).

[79] Che-Wei Chang and Dong Dong. "A Current-Balancing Gate Driver for Dynamic Current Sharing of Paralleled SiC MOSFETs With Kelvin-Source Connection". In: *IEEE Transactions on Power Electronics* (2024).

[80] Ming Zhang et al. "An Improved Methodology for Measuring the Conduction Characteristics of SiC MOSFETs". In: *IEEE Transactions on Instrumentation and Measurement* (2025).

[81] Mingxue Huo et al. "Physics-Informed Intelligent Thermal Model for SiC Power Modules With High Generalization". In: *IEEE Transactions on Power Electronics* (2025).

[82] Zizhao Wang et al. "A Short-Circuit Protection Method for SiC MOSFET Using Combined Vgs and Vds Detection". In: *IEEE Transactions on Power Electronics* (2025).

[83] Yanchao Liu et al. "Auxiliary Commutated Resonant Pole-Based DC–AC Converter With Ultralow THD for Planar Actuation System With SiC Devices". In: *IEEE Transactions on Industrial Informatics* (2026).

[84] Fengtao Yang et al. "Investigation on Short-Circuit Characteristics and Failure Modes of SiC Planar MOSFETs With Linear and Hexagonal Topologies". In: *IEEE Transactions on Device and Materials Reliability* (2025).

[85] Mingxue Huo et al. "Screening Indicator Selection and Clustering Method for Dynamic Current Balancing of Paralleled SiC MOSFETs". In: *IEEE Transactions on Power Electronics* (2025).

[86] Ming Zhang et al. "An Improved Crosstalk Suppression Driver Topology for SiC MOSFET With Fast Switching Transient in the Phase-Leg Configuration". In: *IEEE Journal of Emerging and Selected Topics in Power Electronics* (2026).

[87] Peng Sun et al. "Overcurrent Protection Enabled by Broadband Rogowski Coil Current Sensor for Medium-Voltage SiC MOSFET". In: *IEEE Sensors Journal* (2025).

[88] Mustafa Memon et al. "Mitigation of Machine Neutral Point Overvoltage in SiC Motor Drives by Controlling Antiresonance Frequency and Switching Frequency". In: *IEEE Transactions on Power Electronics* (2025).

[89] Da Zhou et al. "An Accurate Power Loop Stray Inductance Extraction Method for SiC MOSFETs Based on Forced Resonance With Decoupling Capacitor". In: *IEEE Transactions on Power Electronics* (2025).

[90] Tianlong Benjamin Albert et al. "Diagnosing Thermal-Interface Aging of Power Devices Using Self-Sensing". In: *IEEE Transactions on Power Electronics* (2025).

[91] Fei Xu et al. "Modeling and Analysis for Switching Crosstalk of SiC MOSFETs Considering Temperature-Dependent Characteristics". In: *IEEE Transactions on Power Electronics* (2025).

[92] Xiang Zheng et al. "A Novel Level Shifter Driver for SiC MOSFET With Passive Triggered Clamping Circuit". In: *IEEE Transactions on Power Electronics* (2025).

[93] Ajay Rai, Shivam Sharma, and Praveen Kumar. "Active Voltage Balancing of Series-Connected SiC MOSFETs During Body-Diode Turn-Off Using an Active Gate Driver". In: *IEEE Transactions on Power Electronics* (2025).

[94] Cai Chen et al. "A High-Performance 3-D Packaging Structure of SiC Power Module for High-Temperature and High-Power-Density Applications". In: *IEEE Transactions on Power Electronics* (2025).

[95] Qing Li et al. "A Unified Physical PSPICE Model of SiC MOSFET With Automatic Parameter Extraction". In: *IEEE Journal of Emerging and Selected Topics in Power Electronics* (2024).

[96] Ping Wu et al. "Investigation on the Electrical–Thermal–Mechanical Performance of Multichip SiC Power Device With Cu-Clip Interconnect". In: *IEEE Transactions on Power Electronics* (2025).

[97] Christoph H. van der Broeck et al. "A Simple and Physically Insightful Model for Generalized Switching Prediction of SiC MOSFET and GaN HEMT Based Converters". In: *IEEE Transactions on Industry Applications* (2025).

[98] Xiaolei Wang et al. "Bond Wire Failure Monitoring of SiC MOSFETs Based on the Body Diode Voltage". In: *IEEE Transactions on Industrial Electronics* (2025).

[99] Zeyu Liu et al. "Discontinuous Modulation-Based Active Thermal Control for Enhancing the KVA Limit of Grid-Connected Inverter-Based Resources". In: *IEEE Transactions on Power Electronics* (2025).

[100] Peng Sun et al. "Warpage Optimization for SiC MOSFET Multichip Fan-Out Panel-Level Packaging With Thermal–Mechanical Finite-Difference Numerical Modeling and In Situ Digital Image Correlation Validation". In: *IEEE Transactions on Components, Packaging and Manufacturing Technology* (2025).

[101] Cheng Li et al. "SiC MOSFET Solid-State Circuit Breaker-Based Motor Control Center: Design Analysis, Novel Soft Start, and Soft Turn-Off Strategies". In: *IEEE Journal of Emerging and Selected Topics in Power Electronics* (2026).

[102] Jun Wang et al. "Temperature Scaling and C–V Modeling of SiC Low-Voltage MOSFETs for IC Design". In: *IEEE Transactions on Electron Devices* (2025).

[103] Fengtao Yang et al. "Surge Characteristics of Planar-Gate Silicon Carbide MOSFET in the Third Quadrant". In: *IEEE Transactions on Device and Materials Reliability* (2025).

[104] Hu Tan et al. "Novel Hybrid Cross-Connected Submodule-Based MMC for MV Applications". In: *IEEE Open Journal of Power Electronics* (2026).

[105] Mingxue Huo et al. "Case Temperature Waveform Similarity-Based Online Aging Monitoring for SiC MOSFETs of Accelerated Power Cycling Tests for DC-SSPCs". In: *IEEE Transactions on Industrial Electronics* (2025).

[106] Ming Zhang et al. "A Fast Short Circuit Detection Method for SiC MOSFETs Based on Drain Voltage Swing". In: *IEEE Transactions on Power Electronics* (2026).

[107] Hong Chen et al. "A Novel Transient Thermal Characterization System for Power Modules". In: *IEEE Transactions on Instrumentation and Measurement* (2025).

[108] Yibo Gao et al. "A 200 °C SiC Phase-Leg Power Module With Integrated Gate Drivers: Development, Performance Assessment, and Path Forward". In: *IEEE Transactions on Power Electronics* (2026).

[109] Pengju Sun et al. "Integrated Thermal Management for a High-Power-Density Silicon Carbide Power Module With Die-Level Heat Flux Over 1000 W/cm2". In: *IEEE Transactions on Industrial Electronics* (2026).

[110] Zeyu Liu et al. "SiC MOSFETs: The Inevitable Trend for 800V Electric Vehicle Air Conditioning Compressors". In: *IEEE Transactions on Transportation Electrification* (2025).

[111] Jun Wang et al. "Demonstration of Gate-Related Trap Characterization in 4H-SiC MOSFETs Using Gate Stress Leakage Current". In: *IEEE Transactions on Electron Devices* (2025).

[112] Ming Zhang et al. "Displacement Damage and Total Ionizing Dose Induced by 3-MeV Protons in SiC Vertical Power MOSFETs". In: *IEEE Transactions on Nuclear Science* (2025).

[113] Hong Chen et al. "Total-Ionizing-Dose and Displacement Damage Effects in Trench SiC Power MOSFETs". In: *IEEE Transactions on Device and Materials Reliability* (2025).

[114] Hu Tan et al. "A Scalable Isolated Gate Driver With Programmable Frequency and Duty Cycle for Series-Connected SiC MOSFETs". In: *IEEE Transactions on Power Electronics* (2026).

[115] Pengju Sun et al. "Adaptive Balancing of Series-Connected SiC MOSFETs Based on Active Clamping and Comparison Circuitry". In: *IEEE Journal of Emerging and Selected Topics in Power Electronics* (2026).

[116] Ming Zhang et al. "Temperature Dependence of Heavy-Ion-Induced Leakage Current in SiC Power MOSFETs". In: *IEEE Transactions on Nuclear Science* (2025).

[117] Zeyu Liu et al. "Review of Cryogenic Power Electronics for All-Electric Aircraft". In: *IEEE Transactions on Transportation Electrification* (2026).

[118] Jun Wang et al. "An Accurate SiC MOSFET Transient Modeling Method for the FPGA-Based Real-Time Simulation of Power Electronic Converters". In: *IEEE Transactions on Industrial Electronics* (2025).

[119] Cheng Li et al. "Optimizing the Number of Parallel SiC MOSFET Chips in a Wireless Power Transfer Inverter With Pulse Width Control". In: *IEEE Transactions on Power Electronics* (2026).

[120] Yibo Gao et al. "Thermal Stability of Gate Driver Circuits Based on 4H-SiC MOSFETs at 300°C for High-Power Applications". In: *IEEE Journal of Emerging and Selected Topics in Power Electronics* (2025).

[121] Hu Tan et al. "Stacked Strongly Coupled GaN/SiC Cascode Device With Fast Switching and Reclaimed Strong dv/dt Control". In: *IEEE Open Journal of Power Electronics* (2026).

[122] Hong Chen et al. "Proton Energy Dependence of SiC Power MOSFET Single-Event Burnout Sensitivity". In: *IEEE Transactions on Device and Materials Reliability* (2025).

[123] Zeyu Liu et al. "Effects of Current Commutation Methods on SiC MOSFET and Si IGBT Based 4-Quadrant Switches". In: *IEEE Transactions on Industrial Electronics* (2025).

[124] Mustafa Memon et al. "Adaptive Determination of Optimum Switching Frequency in SiC-PWM-Based Motor Drives: A Speed-Dependent Core Loss Correction Approach". In: *IEEE Transactions on Power Electronics* (2026).

[125] Cheng Li et al. "Design of a Low-Latency dv/dt and di/dt Closed-Loop Active Gate Driver for SiC MOSFETs With Simple Structure". In: *IEEE Journal of Emerging and Selected Topics in Power Electronics* (2026).

[126] Ming Zhang et al. "Review on Silicon Carbide-Based High-Fundamental Frequency Inverters for High-Speed Drives". In: *IEEE Transactions on Industrial Informatics* (2026).

[127] Hu Tan et al. "A DC SSCB With Range-Limited Parallel Operation Switches and a Snubber Circuit Utilizing Inverse Current Injection". In: *IEEE Transactions on Power Electronics* (2025).

[128] Jun Wang et al. "Heavy-Ion Microbeam Studies of Single-Event Leakage Current Induced by Long- and Short-Range Particles in SiC Power Devices". In: *IEEE Transactions on Nuclear Science* (2025).

[129] Mingxue Huo et al. "Power Cycling Testing for Power Semiconductor Switches: Methods, Standards, Limitations, and Outlooks". In: *IEEE Open Journal of Power Electronics* (2025).

[130] Hong Chen et al. "An Improved SiC MOSFET With Integrated Schottky Contact Super Barrier Rectifier". In: *IEEE Transactions on Electron Devices* (2025).

[131] Zeyu Liu et al. "Artificial Intelligence-Driven Thermal Management in Electric Vehicle Traction Inverters: Cooling System Optimization and Real-Time Temperature Prediction". In: *IEEE Transactions on Transportation Electrification* (2026).

[132] Mingxue Huo et al. "Hyperbolic Deep Learning System for Intelligent Gate-Driving and Health Monitoring of Silicon-Carbide Power MOSFETs". In: *IEEE Transactions on Industrial Informatics* (2026).

[133] S. Sato et al. "Key technologies supporting high performance and reliability of SiC VMOSFET". In: *IEEE Transactions on Electron Devices* (2025).

[134] Jun Wang et al. "Thermal Analysis and Switching Performance of a 1.7 kV SiC Power MOSFET Under High-Frequency Operation and Double Pulse Testing". In: *IEEE Journal of Emerging and Selected Topics in Power Electronics* (2025).

[135] Cheng Li et al. "Quantitative Analysis of Reliability and Cost Function of 15-Level Inverter". In: *IEEE Open Journal of the Industrial Electronics Society* (2025).

[136] Pengju Sun et al. "Multichip SiC Power Module Packaging Using Direct Ink Writing". In: *IEEE Transactions on Components, Packaging and Manufacturing Technology* (2026).

[137] Hong Chen et al. "Investigation of Packaging Solutions for SiC MOSFET Baseplateless Modules". In: *IEEE Transactions on Power Electronics* (2025).

[138] Ming Zhang et al. "Analysis of Avalanche UIS Ruggedness of Vertical Power Silicon and SiC NPN BJTs". In: *IEEE Transactions on Power Electronics* (2025).

[139] Yibo Gao et al. "Comprehensive Experimental and Simulation Study of Six 1.2 kV SiC MOSFET Layout Topologies". In: *IEEE Transactions on Electron Devices* (2025).

[140] Hu Tan et al. "Development of Online Temperature Monitoring Circuitry for Wide Band-Gap Transistors in Motor Drive Applications". In: *IEEE Transactions on Industrial Electronics* (2026).

[141] Zeyu Liu et al. "High-Accuracy Online Junction Temperature Measurement of Medium-Voltage SiC MOSFETs Based on Passive Turn-On Delay Time Integrator". In: *IEEE Transactions on Power Electronics* (2025).

[142] Jun Wang et al. "An Enhanced Characterization Method of SiC MOSFET in High-Voltage Saturation Region with Reduced Self-heating". In: *IEEE Transactions on Industrial Electronics* (2025).

[143] Cheng Li et al. "Voltage-Driven Reverse Recovery in Semiconductor Devices: A Diffusion Capacitance Perspective and Behavioral Model for SiC MOSFETs". In: *IEEE Transactions on Electron Devices* (2026).

[144] Hu Tan et al. "Single-Gate Driving and Nonisolated Power Supply Technology for Series SiC-MOSFETs in High-Voltage Applications". In: *IEEE Journal of Emerging and Selected Topics in Power Electronics* (2026).

[145] Hong Chen et al. "Gate Voltage Dip as a New Indicator for Online Health Monitoring of SiC MOSFETs". In: *IEEE Transactions on Device and Materials Reliability* (2025).

www.ingramcontent.com/pod-product-compliance
Lightning Source LLC
La Vergne TN
LVHW051054210926
848486LV00002B/112